Soil

Ken Simpson

*Formerly Vice-Principal, East of Scotland
College of Agriculture*

Longman   London and New York

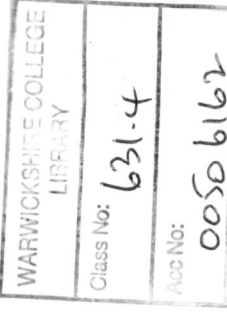

Longman Group Limited
Longman House, Burnt Mill, Harlow
Essex CM20 2JE England

*Published in the United States of America
by Longman Inc., New York*

© Longman Group Limited 1983

First published 1983

Set in 10/11 pt Linotron 202 Times
Printed in Hong Kong by
Astros Printing Ltd

British Library Cataloguing in Publication Data

Simpson, Ken

Soil. – (Longman handbooks in agriculture)
1. Soils – Great Britain
I. Title
631.4′941   S599.4.G7

ISBN 0-582-44641-4

Library of Congress Cataloging in Publication Data

Simpson, Ken, 1921–
Soil.

(Longman handbooks in agriculture)
Bibliography: p.
Includes index.
1. Soil science.   2. Soil management.   3. Soils — Great Britain.
I. Title.   II. Series.
S591.S55   631.4   82–169
ISBN 0-582-44641-4   AACR2

Contents

Preface

We have acquired during the last century a vast store of scientific knowledge about the soil and we have, in the British Isles, very efficient agricultural advisory services to which we can put our problems. However efficient these services may be, they are no substitute for a good basic knowledge of soils, their formation, properties and associated problems. The biggest obstacle to the acquiring of such knowledge is the very wide range, in terms of numbers and ease of understanding, of publications in which the information can be found. The aim of this book is to break down this barrier. 'Jargon' terms have been avoided wherever possible. A separate index of those used has been included so that the user who is not reading the book from start to finish may easily find essential definitions and descriptions.

Apart from my own knowledge of soil gained over the past 40 years in teaching, research, development and advisory work, I have consulted a very wide range of published works. Many of the figures and tables have been constructed from information brought together from several sources. Others are reproduced directly or redrawn from the data of a single published work. These have been acknowledged.

I wish to record my great debt of gratitude to Dr Phil Crooks and to Mr Ron B. Speirs of the Edinburgh School of Agriculture for their considerable and varied help in discussions, supplying much valuable information and in criticizing draft manuscripts.

My thanks go also to Mr Colin Hunter, farmer and former student at the Edinburgh School of Agriculture, for the loan of an excellently edited set of notes taken from my lectures from which all the chaff had been skilfully winnowed.

K. S.
Edinburgh 1981

Soil formation and properties

Part one

Soil – the basis of agriculture

Soil is the basis of agriculture. It is the main raw material from which food is produced. Unlike the raw materials of most industries it can, with supplementation of some of its components, be used again year after year, century after century. Provided that it is used with understanding and care, the farmer has what every industrialist dreams of – a self-perpetuating, self-renovating basis for production. Through the weathering processes and, especially through intense biological activity, nutrients are made available to the plant and the structure of the soil is improved, allowing roots to move freely within it.

Maltreatment of the soil, whether through ignorance, greed or accident, will almost always lead to a deterioration in its 'fertility' – its capacity to produce excellent crops for a very long time. Sometimes the deterioration is catastrophic, as it was in the Dust Bowl of the USA, created by drastic changes in land-use made in ignorance by the settlers. More commonly it is insidious, as in the gradual development of compacted 'pans' in the subsoil caused by the persistent use of poor cultivation techniques.

Deforestation, overcultivation, exhaustive cropping, the unskilled use of farm machinery, failure to maintain lime and nutrient levels in soil can all cause soil deterioration. The failure to maintain or increase the humus content of soils under arable agriculture is more responsible, perhaps, than any other single factor for erosion, deterioration of structure, cultivation problems and reduced water and nutrient availability for crop growth.

The aims of agriculture are to produce the best crops from the soil and the best animals from the crops. To these must be added the all-important aims of soil conservation – to perpetuate and improve production by maintaining or improving the raw material, soil.

Soil is, perhaps, the most complex substance that can be described in one word. It is infinitely variable. In some places it teems with living organisms, in others it is virtually barren. Raw young soils developing in newly deposited wind-borne material contain little or no organic matter while some peats contain 95 per cent. The most acidic soil has a pH value around 3.0 and some alkali soils which contain sodium carbonate have pH values greater than 9.0. Some soils contain so much clay that its cohesive properties cause serious cultivation problems. Others contain so little clay that they rapidly lose plant nutrients in drainage water.

The type of soil inherited by the farmer has been determined by its parent material, the climate, past and present, the 'natural' vegetation that grew on it before it was used for agriculture, the topography – slope, altitude, aspect – and by the recent (in geological terms) actions of man.

The soil that results is the farmer's inheritance. It may be rich land that is easy to work, difficult to injure and well supplied with nutrients. The few who inherit this are lucky indeed. There are problems inherent in the management of almost all soils – problems of drought, erosion, poor drainage, cultivation difficulties, lack of nutrients, acidity, alkalinity. Some problems are unavoidable, like those experienced on chalky soils because of the nature of the parent material. Many other problems are man-induced.

Soils which have remained undisturbed by cultivations, liming, fertilizer treatment and other activities of man for centuries are described in this book as 'natural' soils. Such soils gradually mature. A definite type of vegetation develops and there comes about an equilibrium between soil, climate and vegetation.

The very act of ploughing and preparing these soils for agriculture has brought about radical changes in them. Some of these changes are for the good, others for the bad. The latter must be accepted, to some extent, as some of the hazards of agriculture, but they must also be recognized and many of them can be halted. For the good, we have such practices as land drainage and liming which make possible cropping on a scale otherwise impossible. For the bad we have, all over the world, the loss of vast quantities of organic matter from the soil during the early years after reclamation. If these losses go too far and the humified organic matter content of the soil falls below critical levels many ill-effects can occur, depending on the type of soil. Some light-textured soils become subject to wind erosion, rapid loss of nutrients in drainage water, drought and trace element deficiencies. Low-organic silty soils suffer from breakdown of surface structure resulting in the formation of puddles and 'caps'. Heavy-textured soils suffer worst of all from compaction, pan formation and cultivation difficulties.

Agriculture also makes much heavier demands upon the soil for nutrients than in the 'natural' equilibrium state where losses by leaching and uptake by the vegetation are met by nutrients released through rock weathering, the return and recycling of dead vegetation and animal excreta. The demands of cultivated crops for nutrients, especially in intensive cropping, are many times those of the natural vegetation. For example, a natural pine forest will take from the soil, every year, some 2–3 kg/ha of phosphorus whereas a heavy crop of potatoes will take 20–30 kg/ha.

Lime, manures and fertilizers are applied to the soil to replace these losses and to provide for still greater crop production. Manures, like farmyard manure, are generally rich in many essential nutrients which have come from the soil, and are being effectively recycled. Lime and fertilizers, on the other hand, are an extra input to the soil. Lime is used primarily to reduce soil acidity to such an extent as to allow a range of crops to be grown. Most fertilizers contain, of the essential elements, only nitrogen,

phosphorus and potassium. In the British Isles much of the phosphorus and potassium is imported and the nitrogenous fertilizers are synthesized from atmospheric nitrogen. Their application is a definite gain to offset the increased demands of cropping.

There are, however, at least 20 elements essential to plants or animals or both. Some of them, such as magnesium and sulphur, are required in quite large quantities. Some, called trace elements, including manganese and boron, are required in much smaller amounts. Very little of any of the essential elements, except nitrogen, phosphorus and potassium, is added to the soil in fertilizers. The use of manures, containing a wide range of essential elements is now very restricted in some production systems. In these circumstances, the soil is virtually the only source of magnesium and trace elements from which the plant can draw. Consequently, great strains can be put upon the soil reserves of these elements, especially during periods of rapid growth. Also, liming reduces the availability of most of these elements to the plant, but this must be accepted because of the essential role of lime in crop production. Thus, inevitably, under progressive agriculture, deficiencies of elements other than N, P and K have become more widespread and severe.

These strains on both the physical and chemical properties of the soil are bound to happen as we seek to feed the world by attaining higher and higher crop yields. But it is essential to understand them and, above all, to understand how and why they have arisen and what can be done to minimize them.

Much can be done within the practice of economic agriculture to correct man-made problems, to avoid creating new ones and to increase the capacity of the soil for excellent long-term crop production: that is what this book is about.

Soil formation

2

The solid part of the soil is formed from two broad groups of materials: organic matter and mineral matter. The spaces between solid particles are filled with water and air. Soil organic matter consists of the remains of organisms, both plants and animals, which have once lived on or in the soil, have died, and are partially or wholly decomposed. Rock minerals, modified by many thousands, sometimes millions, of years of weathering make up the remainder, usually the main bulk, of the soil. Some kinds of soil, for example newly deposited alluvium along river banks and soil which has been recently deposited after wind transport, can be almost completely mineral. Other kinds of soil can be almost completely organic, such as the deep peat formations of the Highlands of Scotland and of parts of central Ireland. Between these two extremes are soils with every possible ratio and type of organic and mineral matter. Obviously such widely varying kinds of soil have been formed in many different ways.

Most soils are 'mineral', containing 80 per cent or more of rock-derived material. Table 2.1 covers the main soil-forming minerals,, some of which are primary, having been components of the original crust of the earth. Other minerals are secondary, having been formed from the decomposition products of primary minerals. Some primary and secondary minerals persist in soils and others are decomposed or changed by weathering processes. Table 2.1 gives a guide to the persistence of the minerals

Table 2.1 Soil forming minerals

Mineral group	Composition	Ease of weathering	Contribution to soil
Quartz	Silica	Very resistant	Major part of sands
Felspars (Orthoclase, Plagioclase)	Alumino-silicates	Slow in arid soils, easily weathered elsewhere	Clay minerals, potassium, calcium
Micas: Muscovite–white Biotite–black	Alumino-silicates	Biotite weathers easily. Muscovite less so	Clay minerals, potassium, magnesium, iron
Carbonates: Dolomite, Calcite	Calcium and magnesium carbonates	Easily weathered and leached	Calcium, magnesium. Acidity control.
Amphiboles and Pyroxenes	Alumino-silicates	Easily weathered	Clay minerals, iron, calcium, magnesium
Apatite	Calcium phosphate	Changes to other phosphates	Original source of phosphorus
Iron compounds: Haematite, Limonite	Oxides of iron	Resistant	Colour, clay minerals, iron

in soils and the contributions that they and their decomposition products make to soil properties. Because of its resistance to weathering, in cool humid climates, quartz in the form of sand makes up a large proportion of soils, even those which are described as 'clay soils'.

The inherent fertility of present-day soils has been fundamentally affected by the processes which dominated their formation and by the rock minerals from which they were formed.

The early stages of the formation of mineral soils are dominated by the weathering of rocks. Later, the weathered material may be eroded and transported to new sites where it will form soil. At this stage, other processes, known collectively as soil profile development, deepen, develop and mature the soil until it comes to a virtual equilibrium with its environment.

Rock weathering

Weathering is caused by the powers of nature – rain, frost, sun, the growth of plants, and so on – which combine to soften and break up the rock into smaller and smaller particles which eventually form the mineral parent material of the soil. The early stages of weathering occur on rock faces exposed to the atmosphere and the surface is oxidized and softened. Small crevices appear and particles break off. At this stage the rock may be colonized by mosses or lichens which initiate the simplest form of biological weathering, the results of which can be seen if one tries to remove the lichen and some of the rock comes away with it. Water also percolates into the small crevices and if it freezes, expands, giving rise to pressures which crack or shatter the rock still further (Fig. 2.1).

During the weathering process rock minerals are chemically changed and some of them are totally decomposed. Some of the decomposition products react with each other to form new substances, the most important of which are clay minerals. Other decomposition products are lost from the developing soil in drainage. The type and speed of weathering are greatly affected by climate. For example, in hot wet tropical conditions silica is rapidly weathered and much of it is removed from the soil in drainage water. By contrast, in cool humid climates such as that of the British Isles silica weathers very slowly

Figure 2.1   Rock weathering

and, in the form of sand particles, forms a large part of many kinds of soil. The felspars and micas, on the other hand, and other complex silicates that are primary rock minerals, are relatively easily weathered in cool humid climates. Their weathering releases into the soil plant nutrients such as potassium, calcium, magnesium and iron and also results in clay formation.

The type of weathering, as it is affected by climate, controls the fundamental nature of the mineral part of soil in different parts of the world, from the red tropical soils composed largely of compounds of iron and aluminium to the silica-rich soils of the British Isles.

Weathering can occur by physical, chemical or biological means, and usually by a combination of all three working together.

Physical weathering

Physical weathering occurs mainly under the influence of water, wind and changes of temperature. The result is the progressive breakdown of rock into finer particles.

Water in fast-flowing streams or rivers carries with it abrasive pieces of rock, gravel or sand which wear away rock on the stream bed by rasping off particles. These particles are rolled along by the water and eventually become rounded boulders or pebbles.

Water in the form of massive glaciers during glacial periods, carried with it boulders as grinding tools, and had a strong abrasive action on rocks over which it passed, sometimes grinding the material very finely.

Abrasive particles of sand, carried by wind, wear away the surface of exposed rocks.

Water percolating into rock crevices freezes in winter and expands in doing so to crack the rock still further.

Temperature changes from day to night cause strains to be set up where one mineral joins another within a rock, because the minerals expand and contract to different extents. These strains are very marked in hot desert areas where the differences between day and night temperatures are great and small pebbles can be heard disintegrating.

Tree roots and, to a lesser extent, roots of other plants developing in rock crevices can exert tremendous pressures, sometimes splitting rock to depths of several metres.

Chemical weathering

Four main processes are involved in chemical weathering: solution, hydration, oxidation and hydrolysis. The rock minerals are chemically altered or decomposed.

Solution Few rock minerals are soluble in water, but many of them will dissolve slowly in carbonic acid, H_2CO_3. This acid is formed by the combination of water and carbon dioxide from the atmosphere. Carbonic acid will, for example, convert calcium carbonate, $CaCO_3$, the main component of chalk and limestone into the more soluble calcium bicarbonate, $Ca(HCO_3)_2$. Organic acids, formed during the decomposition of soil organic matter, and mineral acids such as sulphuric acid, formed by oxidation of sulphur compounds, also play a part in dissolving soil minerals.

The main result is the loss from the soil of materials in drainage.

Hydration This is a process by which water is added to the molecule of a chemical compound. A simple example would be the hydration of the rock mineral haematite (ferric oxide Fe_2O_3), so-called because of its blood-red colour, to the weathered mineral limonite (a hydrated ferric oxide $2Fe_2O_3.3H_2O$) which is named for its lemon-yellow colour. The addition of water to the molecule often produces a change in colour and a softening of the material. The hydrated mineral may also occupy a greater volume than the original mineral, and this causes stress within the rock.

Oxidation This process, as it occurs in soil, usually consists of combining oxygen with a chemical compound. This can occur directly by the action of atmospheric oxygen, for example, by the conversion of ferrous oxide, FeO, released by other weathering processes to ferric

oxide, Fe_2O_3, which has a higher ratio of oxygen to iron. (The suffixes '-ous' and '-ic' are used to represent lower and higher oxygen contents). This oxidation of iron compounds results in the reddening of the surfaces of boulders during weathering and also in the reddish brown colours of freely drained soil in which oxygen abounds.

Although direct chemical oxidation undoubtedly occurs, microbial oxidation of compounds is also important and especially concerns iron compounds. The rust-like ferric hydroxide, $Fe(OH)_3$, deposits that are found around the outflow of drains in peaty areas, and similar ochreous deposits, which may actually block drains, are caused by this sort of bacterially induced oxidation.

A more bizarre and remarkable example of oxidation in weathering is the conversion by bacterial action of iron pyrites, FeS_2, to sulphuric acid, H_2SO_4, and ferric hydroxide, $Fe(OH)_3$. Pyrites, which occurs in coal, shale and some slates, is oxidized in this way in the spoil heaps associated with the mining of shale. Water, with dissolved sulphuric acid, percolating into surrounding areas from these heaps renders the soil very acid indeed, resulting in crop failures. It is almost impossible to correct this acidity by liming because of the constant supply of sulphuric acid from the spoil heap. It is a similar process by which the metal guttering on roofs in some parts of Britain is corroded by sulphuric acid derived from otherwise excellent pyrite-bearing roofing slates.

Hydrolysis This process also involves water and is most important in the decomposition of silicate minerals such as felspars and micas. In simple terms, the action of water splits up the silicate into alkaline components such as potassium hydroxide, KOH, and silicic acid, H_2SiO_3. These products do not survive in the soil but are either leached, as silicic acid is in vast quantities in tropical conditions, or react with other substances to form new compounds. The supremely important effect of hydrolysis is the resultant availability of simple silicon compounds for reaction with aluminium oxides to form clay minerals. The type and quantity of clay minerals formed during chemical weathering strongly affect the water holding and nutrient retention properties of soil.

Biological weathering

This is the weathering which occurs under the influence of living organisms. The example given under physical weathering of tree roots cracking rock could be described as biological although it is obviously a physical action. The roots of smaller plants penetrating into rock crevices or into soil cause similar disruptive action. Lichens growing on rock surfaces produce organic acids which erode the rock surface. Also, during the formation of humus, particularly where rainfall is high and parent materials are low in bases, acids are produced which dissolve and cause

leaching of iron and aluminium compounds. Other chemical effects brought about during biological weathering include the bacterial oxidation of iron compounds mentioned under chemical weathering.

Burrowing by vertebrate animals such as rabbits exposes new materials to weathering and can initiate erosion on steep slopes. Within soils which have a high earthworm population, as in well-managed permanent pasture, large amounts of soil are ingested and excreted by earthworms. Some weathering occurs in the gut and the granular structure units of some soils consist simply of the excreta of these animals.

The actions of man as a biological-cum-physical weathering agent cannot be neglected. When armed with a stick of gelignite and supplied with a solid quarry rock face he can further the cause of weathering in a few minutes as much as natural forces could do in thousands of years.

Unfortunately he seldom allows soil to develop subsequently. More universal and significant are man's contributions to weathering by cultivations and, thereby, the repeated exposure of new surfaces to weathering agencies. The catastrophic wind and water erosion in, for example, the United States of America, resulting from the ploughing out of soil unsuitable for arable cropping, are now part of history but the damage done will never be repaired.

It will be appreciated that the results of all the exampes

of biological weathering that have been discussed are either physical or chemical.

Results of weathering

The combination and interaction of physical, chemical and biological forces cause changes in the original rock material varying from a simple reduction in size of individual particles to complete decomposition and sometimes re-synthesis of the breakdown products. In cool humid climates the processes are usually slow and it can be hundreds or thousands of years before soil forms on top of virgin rock. In contrast, weathering is very rapid in some tropical soils involving the leaching of large quantities of silica each year.

The results of weathering may be summarized as:

- the production of smaller and smaller rock fragments;
- softening and increased solubility of the original material by chemical change;
- if oxidizing conditions or silica leaching occur, a reddening of the material due to residual iron oxides;
- loss of potassium, sodium, calcium, magnesium and other plant nutrients;
- loss of silica;

and, most important of all:

- the formation of clay minerals which strongly affect water retention, plant nutrient retention, drainage and the response of the soil to cultivations.

Transported soils

Before or during soil formation the material produced during weathering may be transported by wind, water, ice or man to another site, sometimes a long way from its point of origin as rock or other 'solid geology' formations. All too often the transport takes place by means of catastrophic erosion, for example by flood waters, but the processes can also be slow and gentle as in the case of the gradual 'creep' of soil down a hill slope.

Note that it is usually the parent material which is transported. The way in which this occurs and the nature of the final material which comes to rest and forms soil are critically important in determining the cropping potential of the soil. The resultant material is known as transported soil. A diagram of such a soil is shown in Fig. 2.2. There is a sharp break at point A where the solid geological formation (rock or sedimentary deposits) meets the transported parent material which may not bear any relationship to the basal geology of the area, particularly where the transported material is very thick. Some deposits can be 30 m deep. Often the material from which the soil is formed has come from rocks of several different types and origins. Maps of the solid geology of the area are of little use in identifying transported soil types.

The mode of transport of the parent material is a major factor in determining the kind of soil which will eventually develop: whether the drainage will be good or poor;

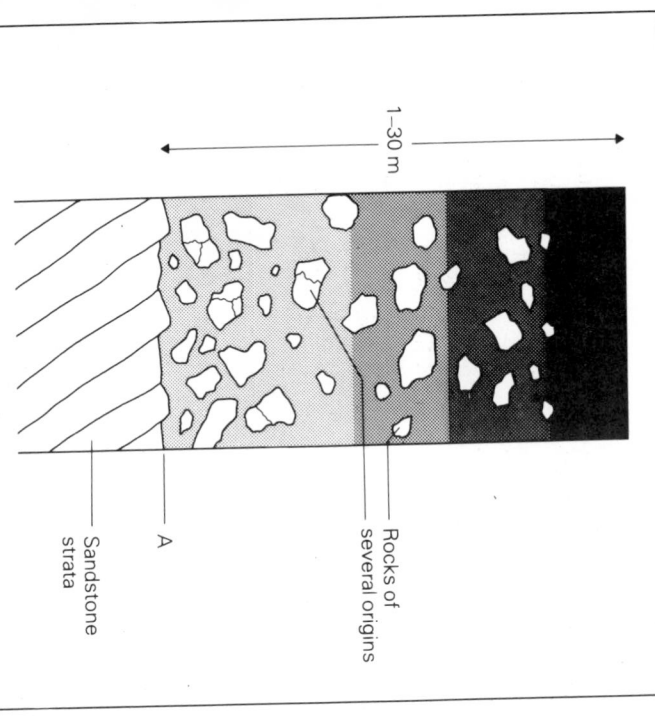

1–30 m

Rocks of
several origins

A

Sandstone
strata

Figure 2.2   A transported soil – glacial till overlying rock strata of
sandstone

whether the chemical fertility will be 'high' or 'low';
whether peat formation is likely to occur.

It is often possible from the topography to identify which
sort of transported soil will be present. For example, in the
flat valley bottoms alongside rivers the soil is usually
alluvial. It is important, however, to recognize that
topography is not a sure guide. In many cases, particularly
on flat or gently undulating country, in which a mantle of
transported material has been deposited by glaciers, the lie
of the land gives few clues to the types of soil to be found
there. Much can, however, be diagnosed about the origin
of the parent material, the way in which it has been
transported and the problems that are likely to be met in
managing the soil, by examining the soil, in the field, to a
depth of about a metre. From a study of location,
topography, soil texture, along with the stone shape, size
and variety of stones, the type of transported soil can be
identified.

Colluvial soils

Colluvial soils are formed from parent material which has
moved down a hill slope under the joint action of water,
frost and gravity either gently over hundreds of years (hill
creep) or vigorously by massive erosive action (scree
slopes). Soil starts to form during the gentle movement of
hill creep soils and continues after the parent material has
come to rest.

These soils are found along the lower slopes of steep hills filling the concave area between the hill and the flat land below. There is usually a marked double change of slope (Fig. 2.3). This feature can often be seen along river valleys bounded by steep hills above the alluvial plain. There is also a remarkable example below Salisbury Crags in the city of Edinburgh where, below the vertical face of the Crags is a slope of debris stretching to the flat land below.

The texture of colluvial soils is usually gravelly or sandy, particularly on higher parts of the colluvial area, due to erosion of finer particles. Lower parts of the slope may have more silt and clay in texture but colluvial soils are seldom heavy textured.

Scree slopes consist of large angular stones of local rock with very little soil material. Hill creep soils have mainly angular stones of various sizes. If glaciation has preceded hill creep there may be a few rounded stones. The amount of stones usually decreases lower down the slope.

Productivity   Scree slopes are useless because of steepness, large stones and lack of fine soil material. Hill creep soils suffer from excess leaching of nutrients and tend to become acid. Slopes are sometimes still quite steep (15–20°) and although the soil may be at the 'angle of rest' under a grass or heather vegetation any disturbance of the plant cover by overgrazing, burrowing of rabbits or inept

heather burning can lead to erosion. On more moderate slopes (11–15°), cultivation can induce a risk of erosion especially if done up-and-down the slope. Other cultivation problems result from the high content of sharp, angular stones. Colluvial soils tend to dry out quickly, because of exposure and rapid percolation of water, and tend to be droughty in summer.

In some colluvial soils on lower, gentler slopes (8–11°), the restrictions mentioned are slight and the soil is fertile.

Alluvial soils

The parent material of alluvial soil is deposited by rivers and is located either where the river in flood has left behind deposits or where the river has changed course. These soils occupy the areas beside streams and rivers (Fig. 2.3.), varying from a few metres in width beside fast-flowing streams to wide areas of alluvium near the river estuaries (estuarine alluvium) of, for example, the Severn, Forth, Trent and Humber. On a much grander scale, the vast areas of soil around the lower Mississippi in the USA and the Ganges in India and Bangladesh are alluvial.

The texture of alluvial soils depends very much on the force of the water by which they have been deposited. If deposited by fast-flowing water, generally near the upper reaches of rivers, the texture is gravelly or sandy and the soil will contain many rounded water-worn stones derived from geological formations along the path of the river. Near the estuaries where water flow is usually slow, fine sand, silt and clay will be deposited and the soil will have a heavy texture. There are fewer and smaller stones in the lower, wider plains and estuarine alluvium is usually stone free.

Productivity Problems arising from periodic flooding are present in all alluvial areas. Water control schemes have reduced the risks on many major rivers but there is always a hazard. The high water-table always within the reach of plant roots has advantages unless it comes too near the surface of the soil. Most rivers have regions of medium-free.

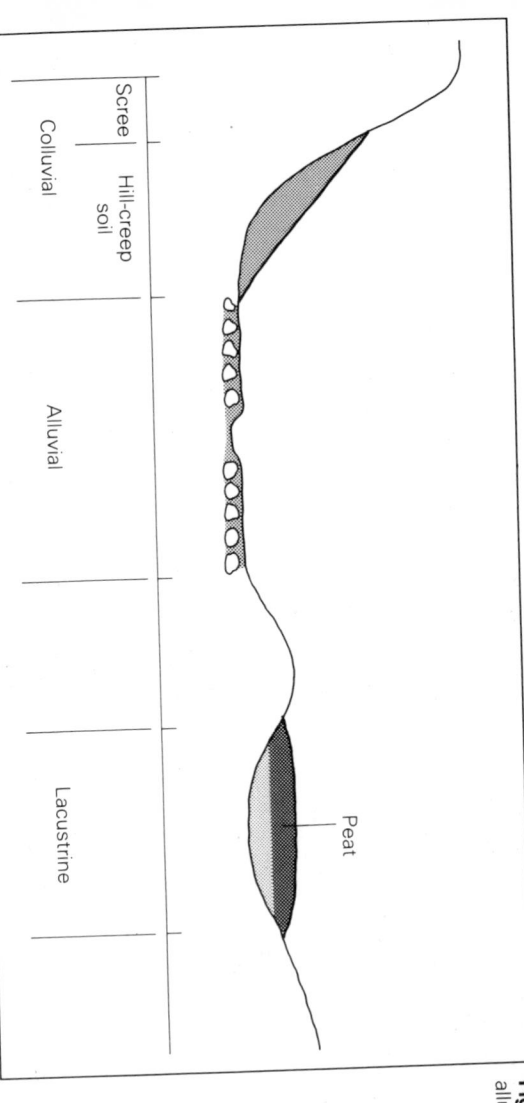

Figure 2.3   Topography of colluvial, alluvial and lacustrine soils

textured alluvium (loams and silty loams) and these give rise to very fertile soils with no drought problems. The silts and clays of estuaries have major drainage problems exacerbated by the flatness of the area. They are usually rich chemically as in them are deposited the fine particles eroded from various geological formations over large land areas. They are, however, difficult to cultivate and slow to warm up in the spring of the year, because of their high water content.

Lacustrine soils

Lacustrine soils are formed from material deposited by very slow-flowing or stagnant water in lakes which eventually become completely sedimented, with the water-table at or near the surface. Because of the very poor drainage, peat tends to form on the surface (Fig. 2.3). The soil is usually stone free and the textures are heavy (silts, silty clays, clay loams, clays) with or without a layer of peat on top.

Productivity The main problem is very poor drainage usually associated with difficulties of outflow if drainage schemes are attempted. If satisfactory drainage can be achieved, especially if a shallow peat of a few centimetres thick is present, the resultant soil can be very productive. Many such areas, however, have developed deep peat

which is difficult to bring into production because of drainage problems.

Glacial soils

The parent material of glacial soils has been transported during the several great glacial periods, stretching over some hundreds of thousands of years, by the ice and by waters from the melting ice at the ends of the glacial periods or during less cold spells within the glacial eras. In some mountainous regions of the world, such as the Alps and the Himalayas, these processes are still active and can be studied in progress.

It is very difficult to imagine the chaos of ice movement and swirling water which gave rise to the massive deposits, some of them 30 m or more in depth, which form one of the major parent materials of soils in the Northern Hemisphere. The result has been a very heterogeneous pattern of soil.

Glacial soil occupies very large areas of the north and west of the British Isles and there are similar glaciated areas in northern Europe and North America.

It is unwise to use topography to predict the presence of glacial soils and, particularly, to differentiate between ice-borne and post-glacial water-borne parent material. The flat or gently undulating terrain of glaciated areas can encompass freely drained sand deposited by water within a few metres of clay loam till deposited by ice.

The hill and mountain tops in most glaciated areas of the British Isles have been rounded as a result of ice action, only a few jagged peaks such as Snowdon and Ben Nevis having been high enough to escape ice cover.

There are three main types of glacial parent material: glacial till, water-worked till, and fluvio-glacial sand and gravel.

Glacial till

This is deposited directly by the ice and is uncovered when it retreats. It can vary in thickness from a few centimetres to 30 m or more. Heterogeneity is the hallmark of this material. The texture can vary from sandy loam to clay, but the great sheets of till which cover much of northerly and westerly parts of the British Isles are predominantly clay, clay loam or sandy clay loam.

Some till is stone free but, characteristically, it contains many boulders mostly rounded or sub-angular and representing each of the geological formations over which the ice has travelled. This has given rise to the term 'boulder clay' used as an alternative by some soil scientists for the term 'till'. Many subsoils derived from till contain strongly weathered boulders of sandstone or shale which have completely softened and disintegrated but are held in their original shape by the surrounding soil.

In some places near the limits of the original ice cover, angular stones of only one or two geological formations are found, which have been carried too small a distance under the ice to become rounded.

Productivity   Soil derived from till has the advantage of containing weathered silt and clay minerals from a large variety of rocks and is, therefore, seldom absolutely deficient in any essential mineral nutrient for plants. There are often problems of poor drainage, caused by the heavy texture, which respond to the installation of drains. Cultivation problems and slow warming up in the spring also result from the heavy texture, especially in areas of high rainfall. This restricts the growing season and the range of crops that can be grown.

Water-worked till

Water-worked till is found where the surface of the original clay till has later been churned up vigorously by water from the melting ice. The clay till contains 40 to 60 per cent of clay, the remainder being silt or sand. Water-worked till contains much less clay and much more sand. The likeliest explanation for this is that free-flowing water has carried away some of the finer silt and clay in suspension either to the sea or to be deposited elsewhere under slow-flowing conditions.

The surface textures of water-worked till are 'medium' (sandy clay loams, loams, sandy loams) and the depth of this layer can be anything from a few centimetres to 1 or 2 m. The material is just as heterogeneous as the original till and has the same range of stones, rounded and sub-angular. The water-worked till is underlain by the original ice-deposited till. Fluvio-glacial sand and gravel, from which almost all silt and clay has been removed, may overlie the water-worked material.

Productivity   Some of the most fertile soils in the country are derived from water-worked till. They combine the advantages of medium texture, good surface drainage and a wide spectrum of plant nutrients in the silt and clay with the presence of a water-table caused by the impervious underlying till. If the water-table is about a metre from the soil surface plant roots can tap it easily. Variations from this very satisfactory set of conditions occur according to the depth at which the original till is found. If there are only a few centimetres of water-worked material drainage problems will occur as they do in the original clay till.

Fluvio-glacial sands and gravels

These materials were deposited by swift-flowing melt water. Presumably they were originally components of till and were swept away and sorted from the finer silt and clay before being redeposited. Deposits vary in thickness from a few centimetres to several metres. Soil textures are gravel, sand, loamy sand or sandy loam. The soils may be very stony (up to 70 per cent stone) or stone free according to the force of the depositing water: the stronger the water flow, the more and larger the stones. Stones are rounded or sub-angular.

Many sand quarries are worked in this material where one can see the stratification and the variety of particle sizes, in a vertical face, including thin bands containing fine sand, silt and even clay, representing quiet periods of tens or hundreds of years under still water.

Productivity   Drainage is excessive and leaching occurs, particularly in high rainfall areas. Even in these areas the soil tends to drought. Nutrients applied in fertilizers are subject to leaching and the soil is inherently low in many nutrients because of the low silt and clay content.

Wind-borne soils (Aeolian)

Wind-borne parent materials cover large areas of the earth's surface, including the great Arabian and north African deserts, sand dunes around many coasts as in the Landes area of the west coast of France where they are 100 m deep. The very large areas of loess in eastern Europe and the states of Iowa, Missouri, Nebraska and

Illinois in the USA are also wind borne. Sand dunes and desert sands contain high proportions of fine sand and shift by the process of saltation, the force of the wind causing particles to bounce along the land surface. Loess, on the other hand, has finer particles and is usually silty in nature. It is a product of dry post-glacial periods and has probably travelled in giant dust storms. In the USA it covers previous soil parent materials, even glacial deposits, to great depths.

The topography of areas of wind-borne soils is either undulating or flat. Predictably, the soils are stone free.

Productivity The coarser types such as dune sand suffer severely from drought and shortage of nutrients such as potassium and some trace elements. Recently deposited wind-borne soils are obviously still subject to wind erosion. The finer silty loess deposits have formed the basis for some of the world's most productive soils in, for example, the state of Iowa. Wind-borne silts are prone to the problem of 'capping' – the formation of a thin layer of very fine material on the soil surface under the influence of raindrop splash. Such layers can prevent seeding emergence.

Raised beach soils

Raised beach parent materials are left behind when the sea

retreats from a coast because of geological changes. They are commonly composed of coarse sand with some fine sand and when they are young usually contain calcium carbonate in the form of broken shells. In some areas, such as the south banks of the River Forth, several flat or gently sloping raised beaches are found, the oldest of which can be 30 m or more above sea level and some hundreds of metres inland from the present coastline. These deposits are post-glacial but there are older deposits, raised in some places more than 100 m above sea level.

Productivity Drainage is excessive and leaching occurs, particularly in high rainfall areas. Even in these areas the soil tends to drought. The soil is inherently low in many nutrients because of the low silt and clay content. With constant applications of organic matter, such as seaweed from nearby coasts, the productivity of the soils can be greatly increased. Any sea salt is rapidly leached from these soils but the sites tend to be exposed to salt spray and on-shore winds. The finer textured raised beaches are also prone to wind erosion. Shelter belts of species such as sea buckthorn (*Hippophae rhamnoides*) are essential. The soils are alkaline and liming is not required until, as in the case of some very old raised beaches, all calcium carbonate is leached.

Man-made soils

A wide range of soils is derived from earth moving, rubbish dumping, spoil heaps, and similar operations, but the most important is that of warping, that is the deliberate flooding of land, usually in estuaries, to allow the artificial build-up of alluvial soil. The enclosure of the polders in the Netherlands, a masterpiece of ingenuity, has created almost a million hectares of new rich land and is the finest example of warping. In the British Isles, warping has been practised only on a small scale along estuaries such as the Mersey and the Humber. It would be fairly simple, if expensive, to follow the Dutch example in the Wash. There is, however, no lack of potentially productive land in the British Isles and it would be easier and cheaper to reclaim the large areas of neglected land in hill and marginal areas or to improve the poor permanent pastures bordering English rivers.

Productivity Warp soils are chemically fertile but present all the problems of drainage shown by lacustrine and estuarine soils.

Sedentary soils

Soils which have formed from parent material which has *not* been transported, but which has weathered and

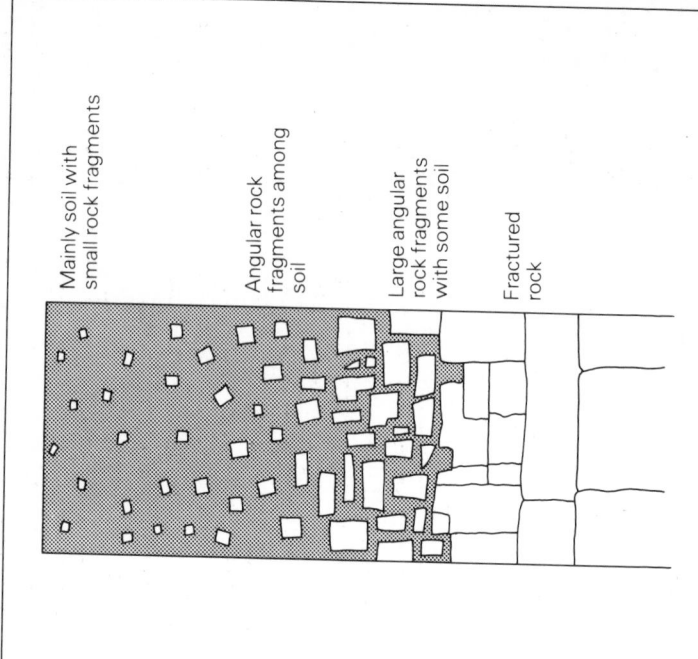

Mainly soil with small rock fragments

Angular rock fragments among soil

Large angular rock fragments with some soil

Fractured rock

Figure 2.4 A sedentary soil developed from massive rock

developed *in situ* may be called sedentary. Figure 2.4 shows a sedentary soil developed from massive rock. In contrast to transported soils there is a direct sequence from massive rock to shattered rock, to finer rock particles mixed with soil, to fine soil material. The parent material is the basal rock of the area and the properties of the soil, while it is shallow and young, are closely related to those of the rock. If weathering has been going on for a very long time and the soil is deep this relationship becomes less close, as a result, for example, of leaching. None the less, the soil will usually retain many characteristics of texture and chemical composition related to the rock. The most obvious examples are soil derived from chalk or sandstone. The former will retain its high pH and the latter its sandy texture for many thousands of years.

There is, therefore, a strong relationship between the basal geology of the area and the kind of soil that develops. Areas covered by sedentary soils are not common in the northerly and westerly parts of the British Isles as the basal geological formations are overlain to various depths by transported materials, mainly glacial. In the Midlands of England there are more sedentary soils and in the south-east they cover large areas, intermingled here and there with transported soils.

The parent materials of sedentary soils fall into four groups: igneous rocks, sedimentary rocks, other sedimentary formations and metamorphic rocks.

Igneous rocks

Igneous rocks were formed 'by the cooling of the earth's crust and have remained unchanged, except at the surface, ever since.

The soils which form directly from them vary according to the texture of the rock and its component minerals. Igneous rocks containing a high proportion of quartz are called 'acidic' rocks. As the proportion of quartz are other minerals to quartz increases they are called 'intermediate' and then 'basic' rocks. Table 2.2 gives examples of igneous rocks according to texture and acidity.

Soils derived directly from igneous rocks do not cover large areas of the British Isles as in many areas the rock has been metamorphosed or, alternatively, has been covered with transported materials. Exceptions to this are the acidic granite-derived soils of the Scottish Highlands and the base-rich basaltic soils of Northern Ireland.

Productivity Fertility of these soils is strongly related to the factors in Table 2.2. Soils derived from coarse-textured acidic rocks such as granite are sandy and easily leached. They are deficient in phosphorus and may be lacking in trace elements such as copper. Coarse-textured basic rocks are more easily weathered to form clays and give rise to fertile, well-structured, near neutral soils of medium-heavy texture. Some fine-textured acidic rocks such as Rhyolite can give rise to freely drained acidic soils with light or

Table 2.2 Some examples of igneous rocks

	Acidic	Intermediate	Basic
% Silica	60–70+	55–60	40–55
% Felspars, micas and other minerals	Less than 30–40	40–45	45–60
Coarse texture (crystals easily visible)	Granite	Syenite Diorite	Gabbro
Fine texture (crystals visible only under lens)	Rhyolite	Trachyte Andesite	Basalt
Mixed texture (large crystals set in a matrix of fine material)		←—— Porphyritic rocks ——→	

consolidated by the tremendous pressures of overlying formations and some have been cemented by silica, calcium carbonate or iron oxides. Limestones and chalk are generally regarded as sedimentary rocks although they would be more accurately described as 'precipitate rocks' because they are formed by the precipitation of calcium and magnesium from solution in the ocean. Sometimes the deposits contain shells, very obvious in some oolitic limestones.

Sedimentary rocks cover large areas of England. Chalk and limestones of varying hardness, from the soft Inferior Oolite to the much harder Carboniferous limestone cover much of south-east and eastern England. There is an area of Dolomite (magnesian limestone) stretching through midland and north-east England. All these materials are harvested for use as liming materials. Sandstones and shales are sedimentary rocks. There are, for example, formations of Bunter and Keuper sandstones which form very sandy soils in the east and west Midlands of England and Carboniferous sandstones and shales in the Pennines. Soils of medium texture are derived from Devonian and Silurian siltstones and from Keuper marl.

Large areas of south-east England are cloaked in very heavy clay deposits which have not been strongly cemented. They are of various ages: Eocene, Triassic, Carboniferous but the most widespread are the Cretaceous and Jurassic clays including the Wealden, Kimmeridge,

medium textures. The most fertile 'igneous' soils are those derived from the base-rich basalts, which are easily weathered to give mildly acid, well structured, medium- or heavy-textured soils.

Sedimentary rocks and other deposits

Sedimentary rocks are the result of weathering, transportation and redeposition of materials in weathering cycles millions of years ago. Many of them have been

Oxford, Lias and Gault. Apart from areas of chalk only small parts of south-east England, such as those occupied by Lower Greensand have light-textured soil.

Productivity   Productivity of these soils is very closely linked to the texture that they have acquired. In the case of limestones and chalk another dominant influence is the very high calcium carbonate content of the soil unless it has been subjected to exceptionally vigorous leaching. At one end of the texture scale, soils derived from sandstones and uncompressed sandy deposits will tend to be acidic and to suffer easily from drought. The heaviest soils derived from unconsolidated clays or from shales will usually be chemically fertile but tend to present drainage and cultivation problems because of their heavy texture. The middle range of textures, derived from siltstones and some marls, give rise to fertile soils with some problems of surface capping.

The soils derived from the limestones or chalk are a class apart, being mainly well drained, well structured, alkaline or neutral but presenting problems of shallowness and drought.

Metamorphic rocks

Metamorphic rocks are formed from either igneous or sedimentary rocks which have been subjected, after formation, to tremendous pressures and high temperatures. They cover large areas of the Highlands of Scotland, parts of north-west England, and Wales.

Metamorphosed igneous rocks include gneisses and schists. Their mineral composition is usually similar to the rocks from which they were formed. Examples of metamorphosed sedimentary rocks are slates formed from shales by tremendous pressure and quartzite derived from sandstone. The former will weather to form clay soils and the latter sandy soils.

Productivity   Soils formed from metamorphic rocks have very similar characteristics to those formed from the igneous or sedimentary rocks from which they have been derived. The main differences are in the speed of weathering which will generally be slower in the metamorphic rocks than in their forbears.

Soil profile development

As soil formation proceeds a soil profile develops and, if sufficient time is allowed without disturbance, matures.

The soil profile is a vertical section of the soil from the surface down to the parent material. It is the grave digger's view of the soil. Some soil profiles are very shallow, no more than a few centimetres, as in the case of a very young soil beginning to develop from rock. Some profiles can be very deep, plant roots may penetrate to depths of some 2 m below the surface.

The whole history of the formation of a soil is summed up in what you can see in the soil profile. It is composed of a number of horizons, which are roughly horizontal layers, one below another, often distinguishable by colour, texture and structure and sometimes only by chemical analysis. The lowest horizon is the parent material.

As soon as the first weathering of rock occurs some soil profile development is initiated. In sedentary soils this will continue uninterrupted but in transported soils, profile development cannot start until the parent material has reached its place of rest. In some cases, as for example a very recent alluvial deposit which I studied following a flood in 1948, the early stages of soil development are rapid. In this case ground flora and trees established rapidly on a small new island because of lack of grazing and there has been a steady development of an organic layer at the soil surface in little over 30 years. The development of a mature soil profile, however, takes many hundreds or thousands of years, during which time it comes to a virtual equilibrium with its environment and becomes stable – provided that it is not disturbed. From the study of such profiles much knowledge can be gained of the parent material of the soil, the way in which it has been formed and of its properties. This information can be put to very good use in classifying and forecasting the problems associated with various groups of soils under agriculture, the top horizons of which have been completely intermixed by cultivations, but the lower horizons are still undisturbed and will affect important functions of the soil such as drainage.

There are two basic processes in soil profile development: movement of substances in water and the incorporation of organic matter.

Movement of substances under the influence of water

In cool and temperate humid climates such as that of the British Isles this usually involves leaching. This is the process of downward movement of substances through the soil under the influence of water. Sometimes the substances are in solid form as in the downward movement of clay particles through soil, and sometimes in solution as in the case of dissolved salts. In semi-arid or arid conditions upward movement can occur as water moves towards the soil surface in the evaporation process.

Incorporation of organic matter

As soon as plants and other organisms begin to live and die in the soil, organic matter is produced, humified and will either build up on the surface of the mineral soil, forming peat, or become mixed with the mineral matter.

The soil profile

The two processes, leaching and organic matter incorporation, help to determine the characteristics of the soil profile and the inherent fertility of the soil.

The visible result seen in the soil profile is the formation of horizons in the soil.

The soil horizons may be briefly described by the use of symbols as shown in Fig. 2.5. The main groups of horizons are:

A – a leached horizon;
B – an enriched horizon;
C – a parent material horizon.

Subscripts are added which may be simply numerical, e.g. A_1, B_2 or, more recently, descriptive, e.g. B_h – a B horizon in which humus has accumulated; B_{Fe} – a B horizon in which iron has accumulated; B_g – a B horizon which is periodically waterlogged.

Factors affecting soil profile development

There are six major factors which by their action and interaction determine the type of soil which develops.

They are parent material, climate, vegetation, topography, time and man.

Parent material strongly influences early profile development through both its chemical properties and its effects on soil texture and drainage. These influences are later modified by the other soil-forming factors but some features such as texture in soils formed from sandstone and alkalinity in soils formed from chalk are very long lasting.

Climate is the most powerful influence on a continental scale, overriding to some extent the influence of parent material to give the major soil groups of the world. The main influences of climate arise from temperature and rainfall which, between them, determine the amount of leaching or upward movement of materials in the soil. Even within an area the size of the British Isles, the effects of climate can be seen in the preponderance of strongly leached soils in the cool, wet north-west, the occurrence of deep peat in poorly drained sites in the same area and the dominance of mildly leached soils in the drier, warmer south and east.

Vegetation, once firmly established under natural conditions, has a strong influence on soil profile development through the type and amount of fresh organic matter it deposits on the soil surface, such as leaf litter in

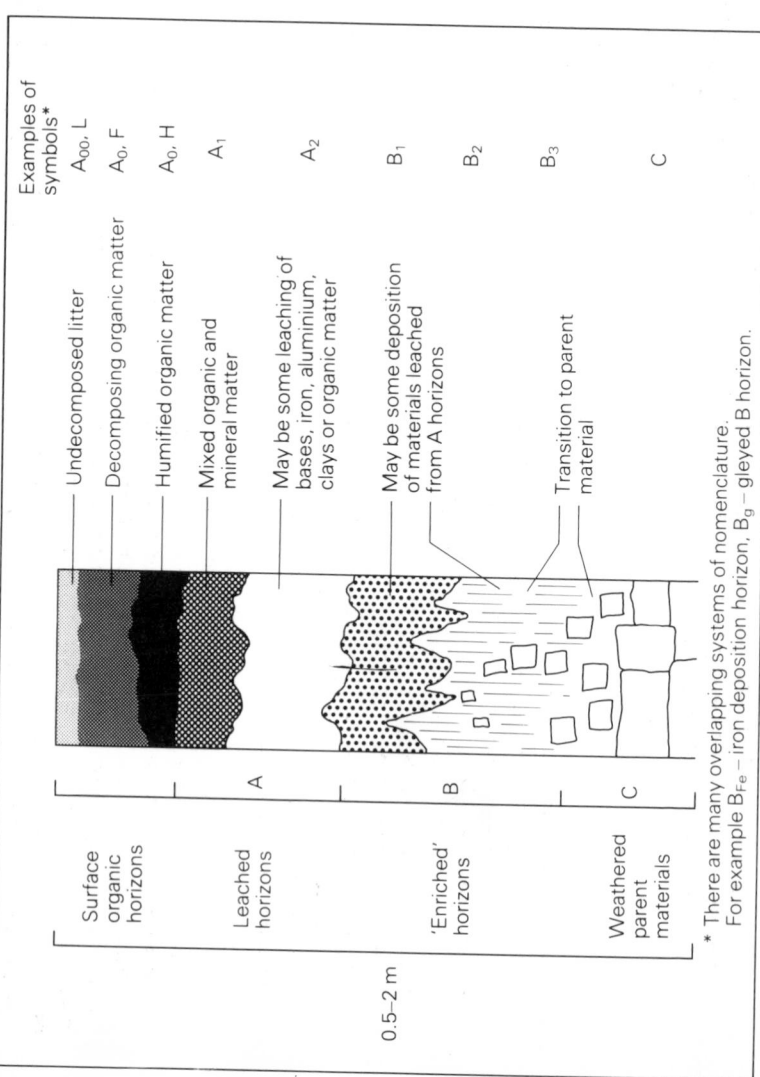

Figure 2.5   The soil profile and its horizons

* There are many overlapping systems of nomenclature.
For example B_{Fe} – iron deposition horizon, B_g – gleyed B horizon.

the forest, or within the soil as root residues. Obviously the type of natural vegetation which can thrive will be dependent upon climate, the humid zones supporting forests which give way to grassland as rainfall decreases and temperature increases across a continent. Vegetation also affects soil development by the transpiration process.

A vigorously growing vegetation transpires large amounts of water and consequently a high proportion of rainfall does not percolate through the soil and leaching is restricted.

As the soil matures the relationship between soil and vegetation becomes very close and the boundaries between plant communities coincide closely with soil boundaries (Fig. 3.5, page 40).

Topography has most of its influence through its effects on local climate. Altitude affects soil temperature and thus the speed of weathering, profile development and the type of vegetation which can grow. Aspect and slope affect the amount of direct sunshine and the amount of rainfall and thus help to control the quantity of water which percolates through the soil. Very striking effects of aspect on vegetation can be seen in central Scotland where acid-tolerant heather dominates the very acid north-facing slopes and grass the less acid south-facing slopes.

Time is required for all aspects of soil profile development. Some processes, such as the production of reddish ferric oxide around roots in soils with fluctuating water-tables can have visual results within months. On the other hand surface organic matter developing under coniferous trees, as seen in the forest species plots in the Forest of Dean, is comparatively slow, only a centimetre or two having accumulated in 100 years. There is also no doubt that centuries or thousands of years are necessary for the development of deep peat.

Man, through agriculture, forestry and other pursuits, greatly modifies the effects of vegetation and topography, by introducing cultivated species of plants, by cultivations and by drainage operations. Also, when virgin soils are first cultivated for agriculture, the surface horizons, the soil from all the A horizons and probably some B horizons, becomes intimately mixed. Only the subsoil horizons remain undisturbed. Man also changes the soil fertility fundamentally for the good by adding lime, manures and fertilizers and often for the bad by poor cultivation practices giving rise to soil compaction, poor drainage and erosion.

Soil formation and soil fertility

Many of the properties of the soil which control or strongly influence its fertility are established during soil formation. They are listed overleaf.

Texture, resulting from the nature of the parent material and the quantity of clay minerals formed during weathering. The texture of the soil is practically unchanged by agricultural practice.

Structure, resulting from the type and quantity of clay minerals and organic matter as influenced by parent material and leaching.

An excellent structure created during soil formation can be ruined in a few years by poor cultivation techniques and over-intensive use. Liming and organic matter maintenance can help to preserve structure.

Organic matter, the type and quantity of which depend on climate, topography, and vegetation.

A reduction of the quantity of organic matter in the soil is inevitable when it is first cultivated. Exhaustive cropping and overcultivation can lead to losses of organic matter and consequent cultivation and erosion problems.

Drainage status, resulting from parent material (texture) and topography.

The natural drainage may be greatly improved by artificial draining and subsoiling. Poor cultivation practices can ruin the drainage by creating pans and sealing off the natural fissures in heavy-textured soils.

Nutrient status, resulting from parent material and climate (leaching).

In agriculture it is normal practice to apply lime and fertilizers containing nitrogen, phosphorus and potassium. Unless organic manures are used extensively there must be a steady depletion of other plant nutrients in the soil.

Thus, of the five main soil properties inherited by the farmer only one, texture, cannot be changed adversely by routine farming practice. Of the others, the most vulnerable are organic matter content and structure which may deteriorate rapidly unless maintenance measures are foremost in the farming system.

Soil groups in the British Isles

3

These are five main groups of soils in the British Isles which have resulted from the complex interactions of soil-forming factors. They are:

Brown Earths
Podsols
Gley Soils
Peat (Organic Soils)
Calcareous Soils

As will be shown, the groups are not clear-cut in the field, but can merge gradually through intergrade soils.

To these main groups must be added Young or Raw soils such as those developing on newly deposited dune sands, on hard rocks or screes. These soils are unimportant in agriculture.

Of the five main groups, the Brown Earths and Podsols are fundamentally well drained and are leached to varying degrees.

Gleys and Organic soils are poorly drained. All four of these groups can be derived from a wide range of parent materials. They account for, between them, more than 80 per cent of agricultural soil in the British Isles.

Unlike the other groups the Calcareous Soils are dominated by their parent materials, chalk or limestone. They occur most widely in south-eastern England.

Many of the characteristics acquired by all five groups during formation are retained when they are used for agriculture, despite the intermixing of the surface horizons by cultivation.

The main characteristics of the groups and their intergrades are described in the next sections.

Brown Earths

Brown Earths are so-called because of the dominantly brown colour throughout the A and B horizons, which gives way gradually to the colour of the parent material in the C horizon. They are leached soils and there is no calcium carbonate in the A and B horizons, although the parent material may contain some.

Figure 3.1 is a diagram of a Brown Earth from the least acid range of the group, derived from a marl containing some calcium carbonate.

This is a very fertile, only mildly acid soil (pH 5.5–6.5) and has been formed under broadleaved forest, usually with an open canopy of oak, ash and a rich herbaceous ground flora. The annual leaf fall from trees and ground flora ensure a constant supply of fresh mildly acid organic matter. Conditions are ideal for earthworms which incorporate the organic matter into the A and B horizons. Humifying bacteria are active. The result is a deeply incorporated humus which is called 'mull'. The structure of the surface horizons is crumb, granular or fine blocky – excellent rounded structure units into which roots can penetrate and in which air and water are well balanced.

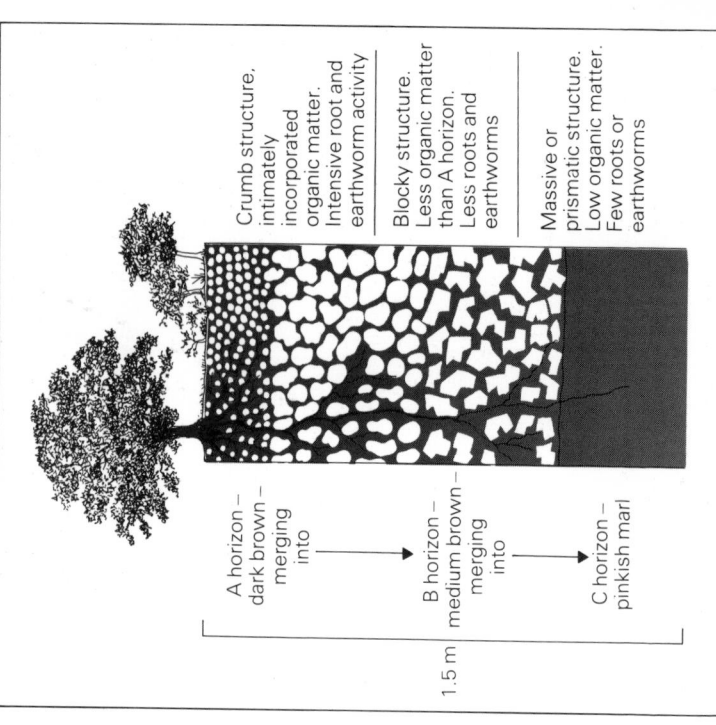

A horizon – dark brown – merging into → B horizon – medium brown – merging into → C horizon – pinkish marl

1.5 m

Crumb structure, intimately incorporated organic matter. Intensive root and earthworm activity

Blocky structure. Less organic matter than A horizon. Less roots and earthworms

Massive or prismatic structure. Low organic matter. Few roots or earthworms

Figure 3.1 A Brown Earth

The A horizon merges gradually with the B horizon which has less organic matter, less roots and less earthworm activity. The structure of this horizon is blocky and the structure units are larger than in the A horizon. The B horizon merges gradually into the parent material (C horizon).

Fertile Brown Earth soils of this kind have largely been reclaimed for agriculture many years ago. They are widespread in areas of moderate rainfall (600–800 mm) and occur on a wide range of base-rich parent materials including shales, marls and clay tills.

The Brown Earth group includes much more strongly leached soils than the one described above. The pH range of the A horizons of the group is from 6.5 to about 4.5. The more acid soils are formed on lighter textured parent materials such as glacial sands and gravels or sandstones in moderate rainfall areas. They also form on more base-rich parent materials in higher rainfall areas (800–1000 mm). The more acid Brown Earth soils have commonly been formed under broadleaved forest including beech and birch, the leaf litter of which is more acid than that of ash, oak and sycamore.

Three main differences from the Brown Earth shown in Fig. 3.1, are found in the acid Brown Earth soils.
● Activity of earthworms is restricted because they cannot tolerate acid conditions. As a result the amount and depth of incorporation of organic matter is less. There may even be a slight build-up of pure organic matter, about a centimetre thick, on the soil surface.
● The structure of the A and B horizons is less satisfactory. Clay is leached from the A horizon and deposited in the B horizon, giving a clay pan which may restrict drainage.
● There are signs of reddening in the B horizon caused by leaching of iron from the A horizon.

This group of acid Brown Earths is also widely used in agriculture. The lowland sandy soils of the group have been cultivated for many centuries. Those occupying marginal or hill land are used largely for stock rearing on grass. All are productive soils under adequate lime and fertilizer regimes.

Podsols

The term 'podsol' is Russian, meaning 'ash-like'. It was used by early Russian soil scientists to describe this soil because of its pale grey leached siliceous horizon underlying the surface organic horizons.

Podsols are strongly leached, very acidic soils with complex profiles. Figure 3.2 shows an example of a member of the Podsol group, the humus–iron Podsol. This is a very acid infertile soil (pH 3.5–4.0) and is commonly formed under close canopied coniferous forest of pine or spruce with no ground flora. It may also be formed under pure heather (*Calluna vulgaris*) communities. Both the

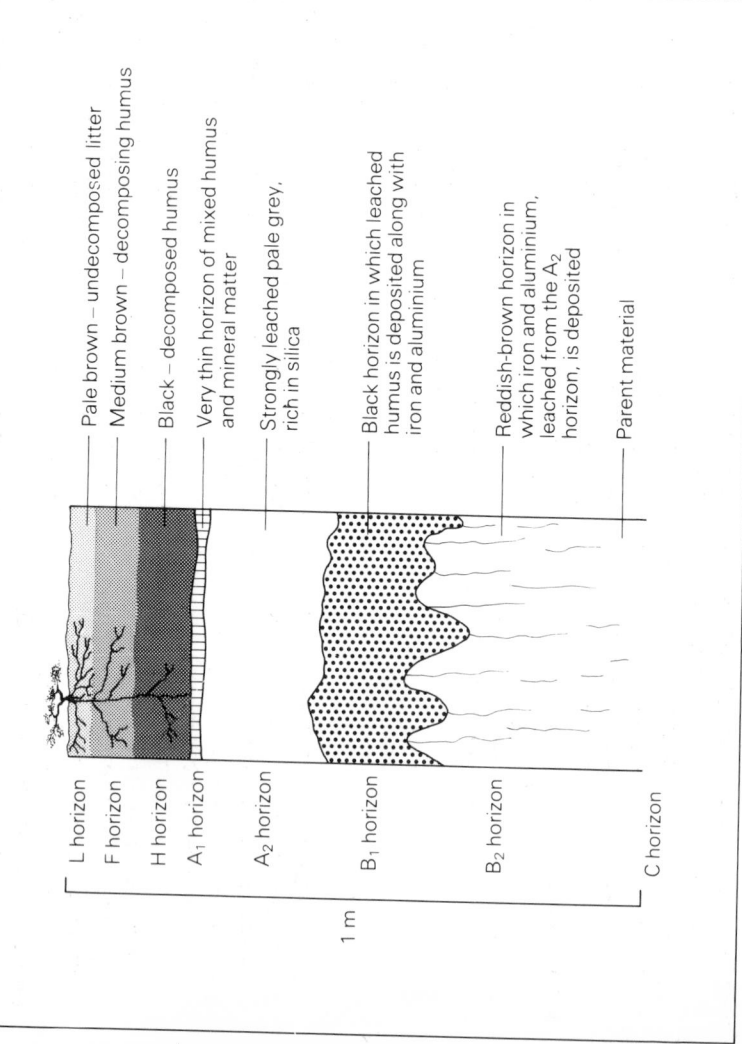

L horizon — Pale brown – undecomposed litter

F horizon — Medium brown – decomposing humus

H horizon — Black – decomposed humus

A_1 horizon — Very thin horizon of mixed humus and mineral matter

A_2 horizon — Strongly leached pale grey, rich in silica

B_1 horizon — Black horizon in which leached humus is deposited along with iron and aluminium

B_2 horizon — Reddish-brown horizon in which iron and aluminium, leached from the A_2 horizon, is deposited

C horizon — Parent material

1 m

Figure 3.2 A humus–iron Podsol

conifers and the heather produce very acid leaf litter.

Podsols develop most strongly on freely drained, light-textured parent materials, low in bases. They occur widely in high rainfall areas (more than 1000 m) on sandstones, quartzite, fluvio-glacial sands or gravels and granites.

Because of the extreme acidity, earthworms are absent and organic matter is not incorporated with the mineral soil. This results in a build-up of organic horizons (L, F and H) on the original surface. These horizons are fully described in Chapter 5. The organic matter is poorly decomposed because conditions are too acid for humifying bacteria. The humus type produced is called 'mor'. It is black and greasy and contains no mineral particles.

The process of podsolization is caused by organic substances, from the organic horizons, combining with iron and aluminium to form soluble compounds which are then leached from the mineral A horizons. This produces the grey E_a leached horizon after which the Podsol is named. In many Podsols everything is leached from this horizon except silica which persists in the form of bleached sand grains.

The organic compounds and some of the 'mobilized' iron and aluminium are deposited in a black B_h horizon. Below this is a reddish horizon (B_{Fe}) in which more iron and aluminium are deposited. This horizon merges with the parent material.

An aspect of podsolization not illustrated in Fig. 3.2 is the formation of a thin iron pan. In some types of Podsol this thin (2–5 mm) strongly cemented horizon rich in iron forms below the E_a horizon. If the pan becomes continuous neither roots nor water can pass through it. The soil, previously well drained, becomes very poorly drained and peat may form at the surface. If this happens the soil is described as a Peaty Gleyed Podsol.

Sandy Podsols in lowland areas are extensively cultivated. They are 'hungry' soils which leach easily and require high lime and fertilizer inputs to produce good crops. Drought is always a risk and crops may fail in dry seasons. Podsols at high elevations can support, at best, poor rough grazing.

Gley soils

The Russian term 'gley' was originally used to describe blue-grey permanently waterlogged clay. It is now used to describe any soil in which drainage is restricted in some way. Gley soils have a zone of water saturation within 2 m of the soil surface. The upper limit of this zone is called the water-table and it normally fluctuates according to season. Thus, compared with the Podsol and Brown Earth, leaching is restricted and air is excluded from the soil.

Gley soils form in meadows on alluvium, especially estuarine, where the water-table is always near the surface.

They also form in lacustrine deposits, and in undisturbed or water-worked till, where a permanent water-table exists because of the impermeable nature of the subsoil. These soils are known as Ground-Water Gleys.

In some cases the water-table is so far below the surface that it does not affect the development of 'normal' A and B horizons, in, for example, a Brown Earth or Podsol, but occasionally reaches the lower B horizons. Soils of this nature are called Gleyed Brown Earths and Gleyed Podsols. If the water-table is near enough to the surface to affect the character of the whole profile the term Ground-water Gley is used.

Many Ground-water Gley soils have been formed as a result of clear-felling of high forests. The consequent reduction in transpiration can raise the water-table by 20–40 cm.

Another type of Gley soil is formed when percolation of water is restricted by a pan in an A or B horizon. This creates a 'perched' temporary water-table and the resultant soil is known as a Surface-Water Gley. Several types of pan can cause this condition. They are described in Chapters 5, 11 and 12.

Ground-Water Gleys

Figure 3.3 illustrates the changes which occur in the soil profile in a sequence of Ground-Water Gley soils in which

drainage becomes progressively poorer. Profile 1 is a well-drained Brown Earth soil. Profile 4 is a soil in which the water-table is always at or near the soil surface and thin peat has developed. This is described as a Peaty Gley. Intermediate conditions are shown in Profile 2 (water-table fluctuating between 1.25 and 0.75 m of the surface) and Profile 3 (0.50–0.20 m). The zone of rise and fall of the water-table is a gley horizon. In the wettest conditions, usually in winter, it is saturated with water. In the driest conditions air moves in as the water-table falls.

The gley horizon is, characteristically, mottled orange-brown to blue-grey. The colours can be very bright in 'meadow soils' alongside rivers, such as the Trent, with a regular rise and fall of the water-table. The orange-brown patches are caused by the formation of rust-like oxidized iron compounds (ferric). It is called ochreous mottling after the reddish haematite and yellowish limonite pigments. The blue-grey colours are caused mainly by reduced iron compounds (ferrous).

In many gley horizons, particularly in lighter soils (sands, sandy loams), the orange-brown mottling takes the form of vertical streaks which are actually tubes of soil rich in oxidized iron, formed around roots. These tubes or cylinders are usually 0.5 to 1 cm in diameter and, in a horizontal section of soil, appear as brown circles in a grey matrix. These tubes are formed because oxidizing conditions occur most frequently around roots as water is

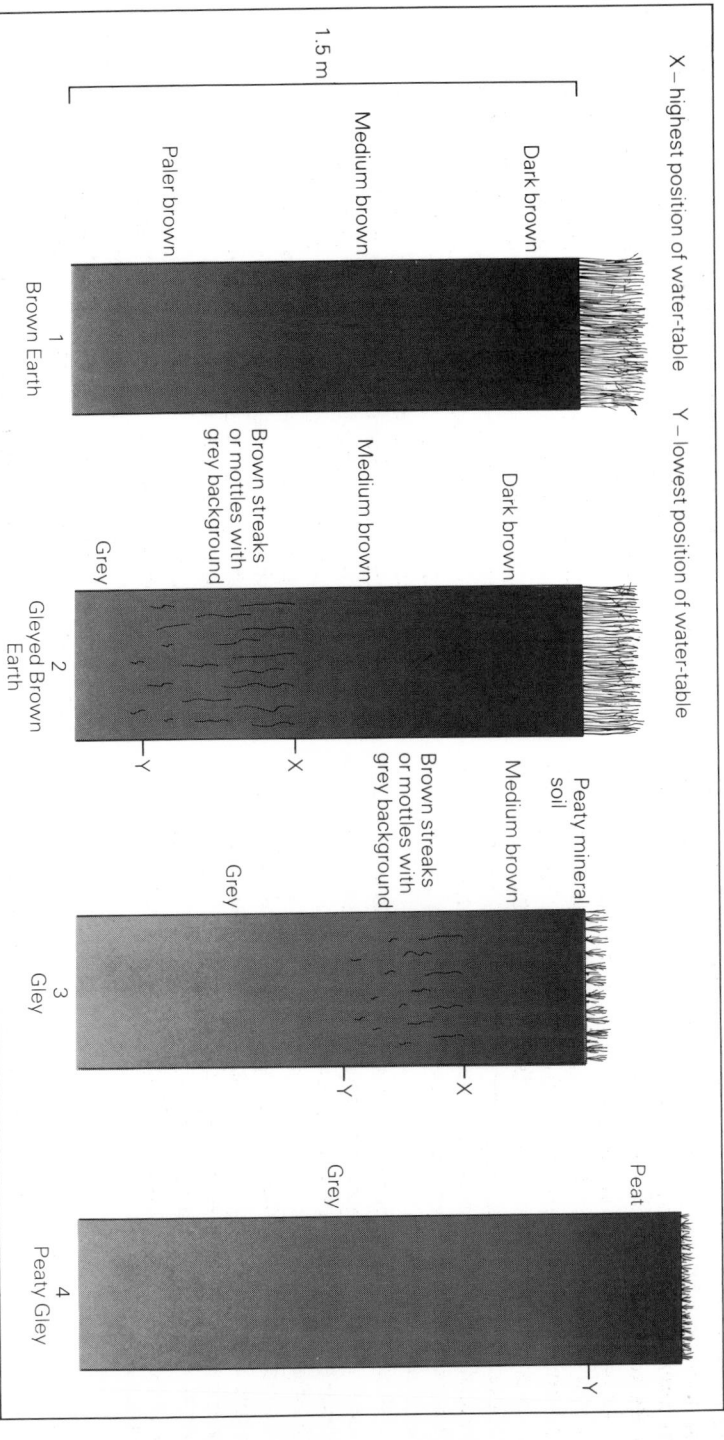

Figure 3.3 A group of Ground-Water Gley soils

drawn from the soil into the plant. They can become quite hard and cemented, around forest tree roots for example, to such an extent that the 'parent' root dies and new roots may penetrate the tube.

In poorly structured, heavy-textured soils (clay loams, clays) there are generally no tubes and the mottling takes on a more angular pattern.

Soil below the lowest level of the water-table will be dominantly blue-grey. The less frequently the soil is water saturated, the more orange-brown colours become dominant and there may be some dark brown streaks identifiable as earthworm channels.

Profile 2 (Fig. 3.3) is a Gleyed Brown Earth and, above the highest level of the water-table will have normal Brown Earth characteristics.

Profile 3 is a Ground-Water Gley. Because of the proximity of the water-table to the soil surface rooting is severely restricted. Earthworms can penetrate below the surface horizons infrequently, the larger burrowing species are absent and incorporation of organic matter by earthworms is, therefore, restricted. The surface soil, although still dominantly mineral, may become peaty in the top 10 cm.

Profile 4 shows a permanently waterlogged soil. Grey colours dominate the mineral soil right up to the surface. The only roots are those of plant species like rushes which can survive in low-oxygen soils. Mixing organisms like earthworms cannot survive, incorporation of organic matter is prevented and peat will build up on the surface.

Surface-Water Gleys

Surface-Water Gleys are widespread in the British Isles, mostly in areas where clay has been leached, under moderately acid conditions, into the B horizons. Several types of parent material are involved, glacial tills and shales being the main ones.

Figure 3.4 illustrates a Surface-Water Gley in which drainage is restricted by a horizon of clay accumulation (clay pan). The fundamental difference between this and the Ground-Water Gley is that the zone of maximum greyness lies immediately above and in the pan. Below the pan drainage is less restricted and the colours tend more towards brown. In Surface-Water Gleys caused by some types of pan (cultivation pans, thin iron pans) this colour change can be quite abrupt and is easily seen.

Many Surface-Water Gleys have two water-tables, one 'perched' above the pan and also a normal ground-water-table. Root penetration of the pan is very poor and because of the very slow upward movement of water in the growing season, the surface soil becomes dry very quickly. The vegetation associated with Gley soils is very varied. If the gley horizons are more than a metre below the surface, the vegetation and the surface horizons will be similar to those of freely drained soils in the vicinity. The true Gley soil with mottling in the A and B horizons is usually associated with rush and sedge communities if the soil is alkaline or slightly acid. More acid Gley soils are likely to develop shallow peat associated with moor mat grass (*Nardus stricta*) and purple moor grass (*Molinia caerulea*).

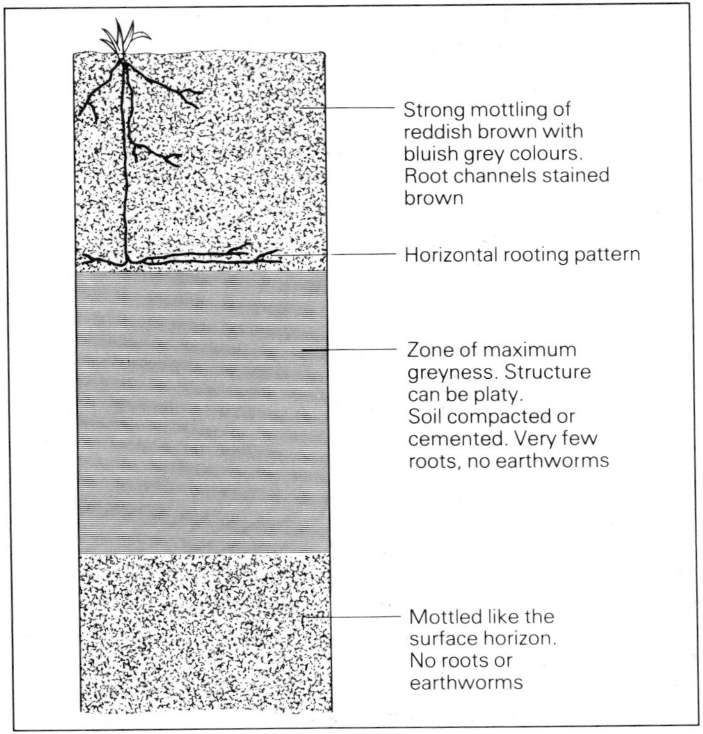
Strong mottling of reddish brown with bluish grey colours. Root channels stained brown
Horizontal rooting pattern
Zone of maximum greyness. Structure can be platy. Soil compacted or cemented. Very few roots, no earthworms
Mottled like the surface horizon. No roots or earthworms

Figure 3.4 A Surface-Water Gley

Peats (organic soils)

Under the conditions shown in Profile 4, Fig. 3.3, peat may continue to build up, depending on wetness and coolness of the climate and some peats are several metres deep, becoming more acidic as they build up. Peat should not be confused with the mor humus of the Podsol. The two build up on top of surface mineral horizons for different fundamental reasons – peat because of poor drainage and mor humus because of acidity, both of which eliminate mixing organisms such as earthworms. Peat is a product of anaerobic conditions whereas mor humus is frequently aerobic. The two are fairly similar in composition but mor humus will never become thicker than 50 cm or so.

Contrary to popular belief, peat is not necessarily acidic. Shallow peat developing on mineral soil rich in bases may be alkaline or neutral. As peat builds up, however, the plants which grow upon it have more and more difficulty in sending down roots to the base-rich mineral matter and gradually acidity develops. Only acid-tolerant species of plants will grow. As they die and decay, the humus formed becomes more acidic and is less well decomposed. Peat will build up to great depths (several metres) only in very wet areas. Some areas, for example north-west Scotland, have such high rainfall that peat develops even on mineral soil in which the drainage is moderate. These may be regarded as climatic peats.

In other areas with lower rainfall peat normally builds up from a lacustrine mineral soil.

The profile of deep peat reflects the succession of vegetation from which it has formed. The peaty mud at or below the original lake surface level contains remnants of semi-aquatic plants. Above these lie the remains of rushes and mosses that colonize the so-called 'flat moor'. Above this can be found heather and other heath plants which replace the flat moor vegetation as the peat builds up, becoming more acid, and rises above the original lake level. It is usually at this stage that trees such as birch, alder and even pine can become established and the transpiration from them can dry out and stabilize the peat and prevent further growth. If trees do not establish, or have grown and have been felled, the peat may continue to grow, becoming very acid (pH 3.0–4.0) and supporting only such species as sphagnum moss and cotton grass. The water-holding capacity of these very acid pale-coloured peats is so high that they defy all attempts to drain them and even tree planting on upturned turfs will usually fail to dry them out because there is a permanent check to tree growth.

The value of peat soils to agriculture is very variable and depends upon site. Shallow peats in relatively dry areas are among the most valuable agricultural soil types provided that drainage problems can be overcome. Relatively deep peats of 1–2 m in thickness usually develop in high rainfall areas and can, with drainage, liming and fertilizer input, give useful pasture and occasionally arable land but the deep peats are very difficult and uneconomic to handle in agriculture and are usually allocated to grouse moor, forestry or recreation. Some are also suitable for cutting as fuel or horticultural peats. Cutting for both purposes has been extensively and successfully practised in Eire.

Calcareous soils

Calcareous soils are derived from parent materials which contain a high proportion of calcium or magnesium carbonate. The main parent materials are chalk and a variety of limestones but there are also sedimentary clays such as the Lower Lias which can be highly calcareous. The soils are alkaline, neutral or very slightly acid at the surface (pH 6.5–8.0).

Calcareous soils cover large areas of southern and eastern England but only limited areas in the remainder of the British Isles. There are two main types: the Rendzina and the Brown Calcareous soil.

The Rendzina is a very shallow dark brown or black soil with a strong granular structure, resting directly on limestone or chalk. The vegetation consists of lime-loving grasses and herbs. Rendzinas are very prone to drought and are of little value in agriculture. They are typical of the Chalk Downs.

The Brown Calcareous soils are deeper than the Rendzinas having 25 cm or more of topsoil. They tend to be silty in texture and well drained. Brown Calcareous soils are widespread on limestones such as Oolite. They make fairly good agricultural soils.

A less well-drained variant of the Brown Calcareous soil is formed on Lower Lias clays. It is gleyed at depths of 25 to 60 cm and forms very useful clay soils in the Vale of Evesham. Despite its heavy clay texture it has a very stable structure and carries excellent fruit crops.

Intergrade soils

In Fig. 3.1–3.4 the different types of soil have deliberately been isolated as individual profiles. Soil in reality stretches continuously across an island or continent. In fact, there is an unbroken tract of land from Finisterre to Vladivostok, over which soil types must either merge gradually, with diffuse boundaries, or change abruptly. Both types of change occur.

Figure 3.3 could, more realistically, have been drawn as a continuous diagram with horizons rising or falling, appearing or disappearing as the soils merge. Profiles 2, 3 and 4 in Fig. 3.3 are intergrade soils between Brown Earth and Peat. They may be named: Gleyed Brown Earth, Ground-Water Gley, and Peaty Gley – the final stage of the transition being Peat. This transition depends upon dryness or wetness of the profiles.

Transitions also occur between Podsol and Peat, viz. Podsol, Gley Podsol (gleyed in B horizons), Peaty Gley Podsol (gleyed in A horizons), Peat.

As a simple example of the merging of soils and the intergrades between the main groups Fig. 3.5 shows a block diagram of an east to west valley in south-east Scotland. Records kept on the site shown in this diagram for 15 years gave average rainfall of 1000 mm at 380 m. The altitude at the highest points of the hills is 450 m and the lowest point 330 m. Only the vegetation and organic matter horizons are illustrated in the diagram (not to scale), and other soil properties are given below the diagram.

There is a clear combined effect of parent material and topography, the Peat and Gley soils being derived from

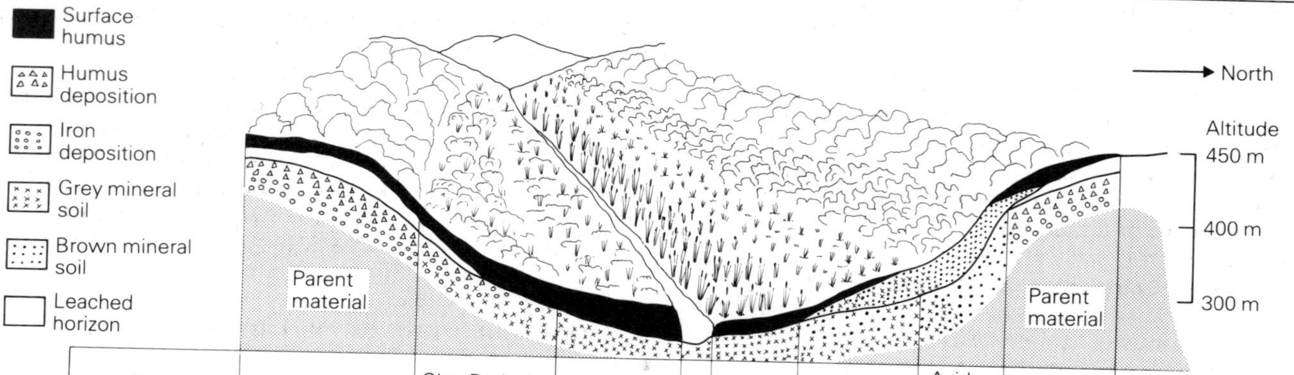

Figure 3.5   Soil transitions across a valley landscape

Soil type	Podsol	Gley Podsol – Peaty Podsol	Peat	Peaty Gley	Gley	Acid Brown Earth	Podsol
Vegetation	Heather	Heather, cross-leaved heath, polytrichum moss	Purple bent, sphagnum moss, cross-leaved heath	Rushes	Moor mat grass	Bents/ fescues, gorse	Heather, blueberry
Surface soil pH	4.0	4.5	4.5	5.0	5.2	5.5	4.5
Humus type	Mor	Transition Mor → Peat	Peat	Shallow peat	Peaty mineral	Mull	Mor
Parent material	Rhyolite	Rhyolite – glacial till	Glacial till	Glacial till	Glacial till	Rhyolite	Rhyolite

Legend:
- ◼ Surface humus
- △ Humus deposition
- ○ Iron deposition
- × Grey mineral soil
- ∴ Brown mineral soil
- ☐ Leached horizon

North

Altitude
450 m
400 m
300 m

Parent material

the lower lying heavy-textured till and the leached Podsols and Brown Forest soils being derived from the coarse-textured Rhyolite.

The effects of aspect and, hence, local climate are seen in the dominance of heather underlain by a Podsol on the cool, wet north-facing slope and the bent/fescue grass community growing on an acid Brown Earth, on the warmer, drier south-facing slope where there is less leaching. On this south-facing slope heather becomes dominant and a Podsol forms only at the highest levels, some 100 m higher than on the north-facing slope.

There is, on this site, a very close relationship between plant community and soil and the boundaries virtually coincide. The boundaries between soils are fairly clear and can be defined within a space of 2–10 m. There are two exceptions. One occurs between the acid Brown Earth and Podsol on the south-facing slope, where a gradual incursion of heather and blueberry mixed with sheep's fescue takes place with increasing altitude over a distance of 30–40 m before heather becomes dominant. An intergrade soil is formed as mor humus gradually builds up on the surface and a leached horizon appears. The other gradual transition occurs towards the foot of the north-facing slope where, over a distance of 50 m the mor humus become greasier and thicker in transition to Peat, the typical Podsol profile develops gley colours in the B horizons (Gley Podsol) and then in the A horizons (Peaty

Gley Podsol) and at the foot of the slope peat is underlain by grey mineral soil, always waterlogged. The vegetation changes reflect this transition with the peat species gradually taking over from the heather.

The relationship between natural vegetation and soil is not always so close as at this site. It depends on the gradual establishment of a stable soil/vegetation equilibrium as the soil matures. There are certain fairly reliable conclusions that can be made from established plant communities especially if they have lasted for a long time.

A few examples are:

bracken (*Pteridium aquilinum*) – Brown Earth, well drained.

gorse (*Ulex europeas*) – Brown Earth, well drained, dry.

rushes (*Juncus* spp.) – Gley soil, usually not peaty.

heather (*Calluna vulgaris*) – Podsol or Peat.

heather/blueberry (*Vaccinium myrtillus*) – Podsol.

heather/bog myrtle (*Myrica gale*)/cross-leaved heath (*Erica tetralix*)/sphagnum moss – Peat.

bent (*Agrostis* spp.)/fescue (*Festuca* spp.) – Brown Earth, well drained.

marjoram (*Origanum*), thyme (*Thymus*), calamint (*Calamintha*) – dry soils based on chalk or limestone (Rendzina).

purple moor grass (*Molinia caerulea*) – Shallow Peat.

moor mat grass (*Nardus stricta*) – Shallow Peat or Gley.

Soil classification and soil maps

4

Soil classification on a continental scale began in the early part of this century when the Russians and Americans elucidated the close relationships between climate, natural vegetation and soil development. Since that time, large areas of the world have been surveyed and mapped on scales varying from 1 : 5 000 000 which can encompass a whole continent to 1 : 10 000 which will show individual fields quite clearly.

Soil classification systems are closely similar to the much further developed botanical classification, there being several levels of classification. The highest level of classification which, strangely, is forgotten by the classifiers, has only one member – soil. This is divided into several groups, often called Orders, selected logically on the basis of major soil properties, such as organic matter type, or soil-forming factors such as climate. These are divided further according to similarity of soil properties until, at the fifth, sixth or seventh level, the units of classification have local significance in agriculture. Each level of classification is named, for example, Soil (usually omitted), Order, Sub-Order, Great Group, Sub-Group, Family or Association, Series, Type. In all classifications a group containing various types of podsol, obviously genetically similar, would be separated out at the second or third level and a Peaty Gley Podsol derived from glacial till containing mainly carboniferous rocks would be identified at the Series level, usually by a proper name often that of a town or village.

The upper levels of classification

A simple example of the criteria used for separating soils at the higher levels is given in Fig. 4.1. It is based on the older USA classification in which zonal soils (1st level) were split according to groups based on climate/vegetation (2nd level). Although it has now been superseded and the terms used for great soil groups discarded, this diagram shows the exciting logic of the system.

It represents a chain of soils from the hot dry south-west of the USA (Arizona, New Mexico) to the cool, wet far north-east (Maine, Vermont, New Hampshire). The classification depends upon the balance between precipitation, P (rainfall + snowfall) and evapo-transpiration, E (potential evaporation losses from soil and vegetation + transpiration by plants). These are equally balanced in the Chernozem and those to the right are increasingly leached as P becomes much greater than E. The profiles to the left become increasingly arid. The natural vegetation is controlled by the climate. The desert scrub in the arid soils giving way to short and then to lush tall grasses as more water becomes available and then to broadleaved trees and finally coniferous forest as leaching

becomes intense. The degree of leaching is graphically illustrated by the position in the profile of the zone of calcium carbonate accumulation. This is at the surface in the arid soils and gradually appears lower down the profile, as the upward pull of evaporating water is diminished and leaching takes over. The strongly leached soils at the right contain no calcium carbonate. The four left-hand profiles in Fig. 4.1 are base saturated even in the surface soil and have pH values of 7 or more. The Prairie soil is leached at the surface and will usually have a pH of 6–7 increasing with depth. Surface pH values of Brown Earth and Podsol profiles are lower and in the Podsol may be as low as 3.5, the leaching being enhanced by the acid litter from the coniferous vegetation.

Further criteria in the classification are the type, quantity and distribution of organic matter. The scant organic matter in the arid soils resulting from the sparse vegetation gradually increases in quantity and deepens until the Chernozem soils have rich, deep, black humus intimately mixed with the mineral matter to the depth of the calcium carbonate layer a metre or more below the surface. The key to the deep organic matter in these soils is the intensive, deep, ramified root systems of the tall grasses. Roots are continually dying, being sloughed off and humified by vigorous bacterial activity in the summer months, and being immediately replaced by new roots. Thus the bulk of the organic matter is deposited and

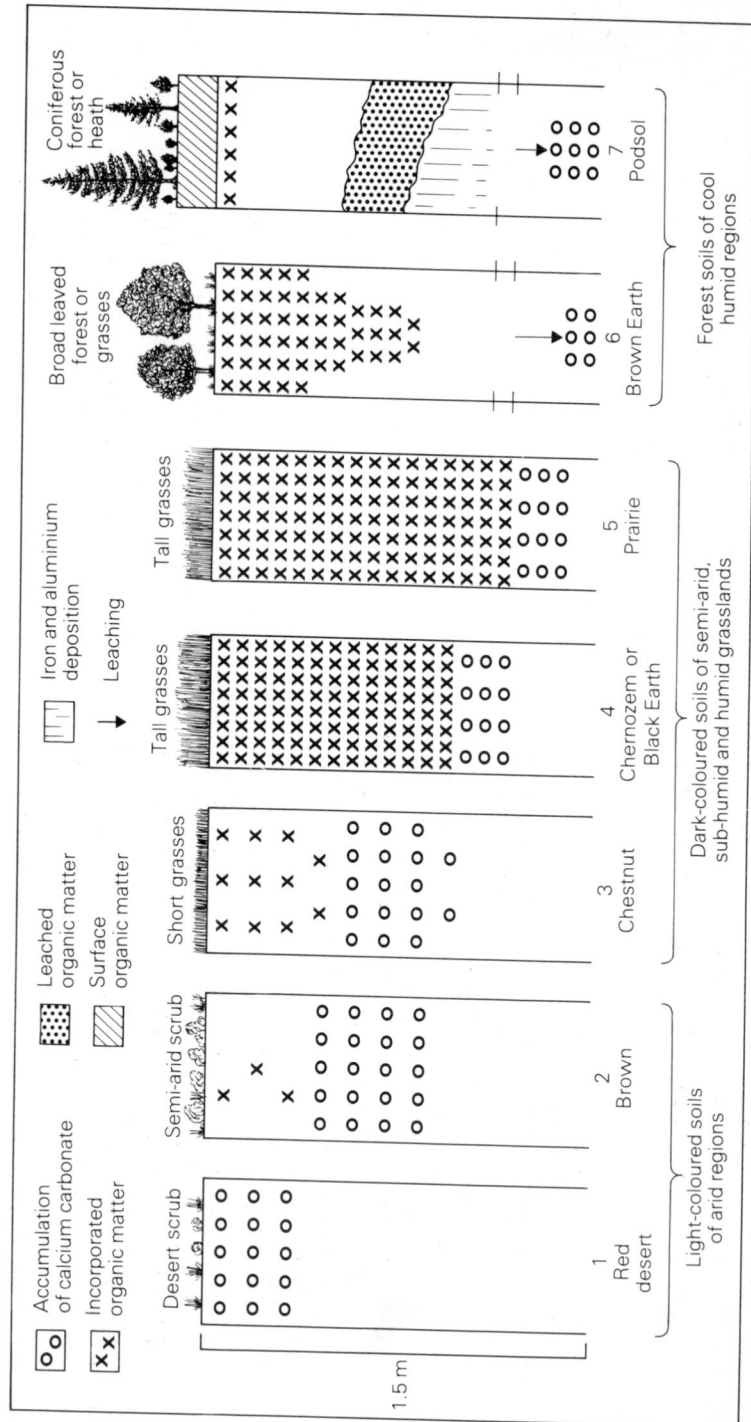

Figure 4.1 Soil variation in the United States of America – a basis for classification

humified within the soil and does not require incorporation.

As we move to cooler and wetter climates the depth of incorporation and the quantity of organic matter decrease. In the forest soils most of the organic matter is deposited on the surface as leaf fall, in contrast with the Chernozem, and must be incorporated by earthworms to form the mull humus of the Brown Earth. Because of the extreme acidity of the Podsol, earthworms cannot survive and mor humus builds up on the surface.

Agricultural potential of the great soil groups

The agriculture and the potential crop yields of the groups shown in Fig. 4.1 vary greatly. The shallow dry Red Desert and Brown soils (note the unfortunate clash of names – this is *not* the Brown Earth) can be rendered productive only by irrigation on a scale that would not be feasible in the British Isles. With irrigation, applied at rates as high as 250 cm of water per year, good crops of citrus fruits and market garden produce can be grown, but costs of production are very high.

The Chestnut soils, east of the Rockies, can be cropped extensively on a wheat/fallow rotation, using farm units of a 1000 ha or more, with low labour input. In such a non-irrigated system yields would be only 1.2–2.0 t/ha of wheat every other year. Alternatively they will produce excellent crops of sorghum, under irrigation, but costs of water are high.

The peak of fertility, as expressed by potential agricultural production, is in the deep, rich, high-humus, well-structured Chernozem on which it is possible to produce yields of 10–12 t/ha of maize (corn) and equally vigorous crops of lucerne (alfalfa). The wetter Prairie soils can do equally well especially in dry seasons when insufficient water may be available in the Chernozem.

None of the soils discussed so far occurs in the British Isles, which are too wet and cool for them to form. Brown Earths are the most highly productive naturally well-drained British soils. On them it is possible to grow, at best, some 10 t/ha of wheat or 50 t/ha of potatoes, although yields generally fall well below this.

Podsols are also widely used in British agriculture and can, after major reclamation problems, sometimes crop as well as adjacent Brown Earths. In general, though, the high rainfall, and light textures of the upper horizons do restrict productivity and the necessary input in terms of lime and fertilizer is high compared with Brown Earths.

Classification and mapping in the British Isles

Many areas of the British Isles have been surveyed and mapped by the Soil Surveys of England and Wales,

Scotland, and Ireland, working separately.

The systems of classification used in the United Kingdom are based on similarities and dissimilarities in the soil profile.

Table 4.1 illustrates the Scottish classification and introduces the terms association and series. The soil association is based on parent material. It may consist of only one or several soil series which, within the parent material grouping are classified by major soil group or sub-group with strong emphasis on drainage class. Table 4.1 shows, in columns, four soil associations from south-east Scotland and, in rows, the series within these associations.

Both associations and series are commonly named after the places where they were first identified. This may help locally, but no picture of the soil can be conjured up by the uninitiated from such names. In the USA they do go a step further by using terms which indicate the soil texture – 'Miami silt loam'. There is a good case for using, in this

Table 4.1 Soil classification in Scotland (from Soil Survey of Scotland. H.M.S.O.)

Association	Parent material	Brown Earths		Iron Podsols	Peaty Podsols	Gleys	Peaty Gleys
Ettrick	Silurian and Ordovician greywackes and shales	Linhope LP	Kedslie KZ	Minchmoor MM	Dod DO	Ettrick ER	Hardlee HR
Winton	Till derived from carboniferous sediments		Winton WN				
	As above with waterworked upper horizons		Macmerry ME			Butterdean BT	
Darvel	Fluvio-glacial sands and gravels derived mainly from lower carboniferous rocks	Darvel DV	Duncrahill DU				
Lauder	Upper Old Red Sandstone Conglomerates	Lauder LA	Spott SO	Langtonlees LL	Ewelairs EW	Lylestone LY	Wakeney WK
Drainage		Free	Imperfect	Free	Free below B₁	Poor	Very poor

country, more cumbersome but interpretable terms such as 'Darvel sandy loam, fluvioglacial podsol'.

The soil association concept has, unfortunately, been used with different meanings by the various Soil Survey organizations. It is of little value to the map user interested in the agricultural value of the soil as it can, for example, embrace soil series having the whole range of drainage classes. The soil series is, therefore, the kingpin of the British classification systems. It is well defined, having a specific type of parent material, falling within a particular major soil group and sub-group and having a definite drainage status.

British Soil Survey maps are based on soil profile examinations, backed up by physical and chemical analysis of horizon samples and by observations of the natural vegetation and topography, both in the field and from aerial photographs. The published maps are on the scale of 1 inch to the mile (1 : 63 000) or later 1 : 50 000. This does not allow identification of average sized fields but the original mapping was done on a scale of approximately 1 : 25 000, giving more detail, and these field maps can usually be consulted at Soil Survey offices. They may indicate, for example, areas of other soils (impurities) occurring within an area mapped as a particular series, but too small to be identified on the published maps.

Figure 4.2 shows a section of a typical soil map but using symbols instead of colours to identify the units. In the published maps colour schemes are devised to indicate the major soil group to which the soil belongs, for example Brown Earths may be shown in shades of brown and orange, Podsols in reddish colours, Gleys in blue, Peaty Gleys in green and Peat in purple.

The Soil Survey, in devising this system, have obviously and correctly placed major emphasis on the processes of soil formation, by putting all cultivated soil into groups which can be classified with certainty only from a study of undisturbed virgin soil. These are extrapolated, skilfully, largely by the use of subsoil characteristics.

A memoir is usually published, along with the map, giving detailed technical information about each soil series and including chapters on the natural vegetation, forestry and agriculture of the area. More recently, simplified versions have been produced for some areas of England and Wales (Soil Survey Records), which are more useful to the non-specialist.

The basic soil map synthesizes information on the parent material, history and present state of the soil and imposes a logical classification to delineate soil series. From these maps, taken along with Chapters 2 and 3 it should be possible to visualize what the soil of a given series will be like and to appreciate the problems it may present. The map may also be used as a starting point for the diagnosis of soil physical problems by examination of individual soil profiles as described in Chapter 9.

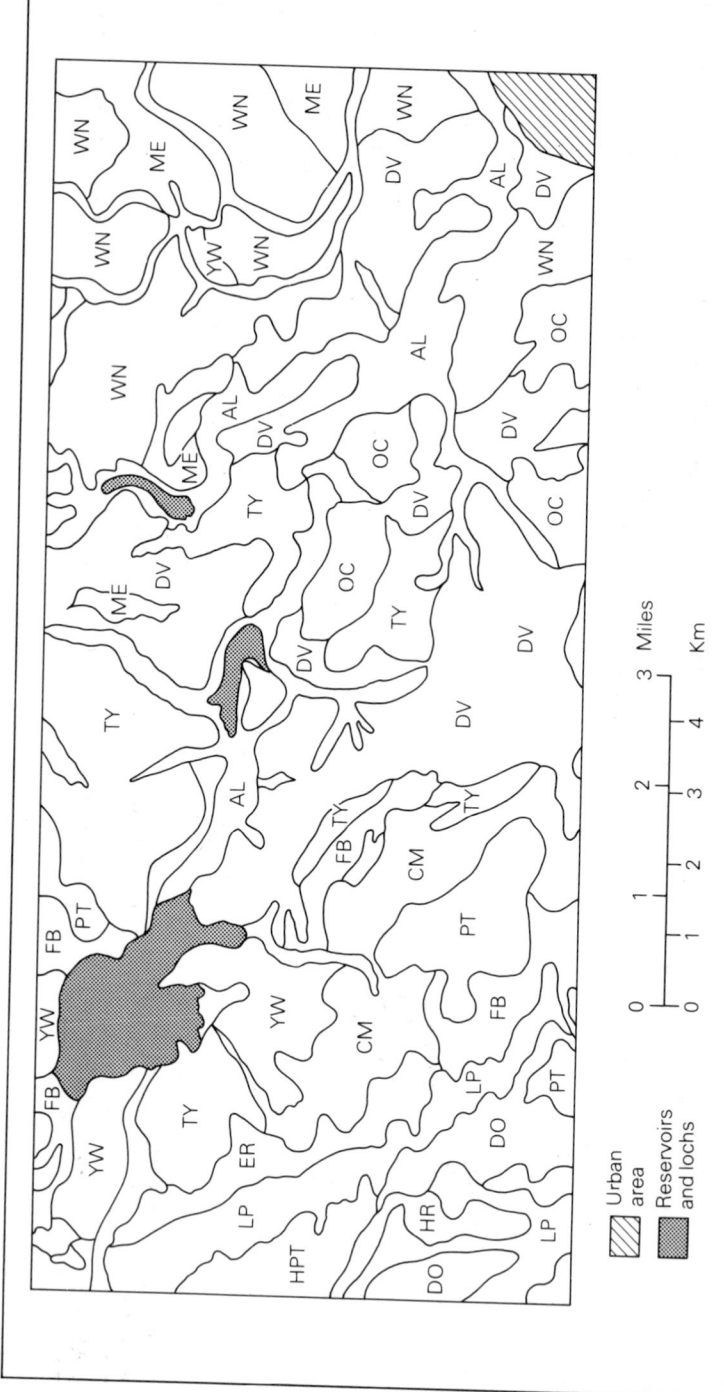

Figure 4.2 A basic soil survey map of part of south-east Scotland (see acknowledgment on facing page)

Association	Parent material	Brown Forest soils Freely drained	Brown Forest soils Imperfectly drained	Peaty Podsols Freely drained below iron pan	Non-Calcareous Gleys Imperfectly drained	Non-Calcareous Gleys Poorly drained	Peaty Gleys Very poorly drained
Darvel	Fluvio-glacial brown sand and gravel derived mainly from Carboniferous sediments	DV Darvel					
Ettrick	Silurian and Ordovician greywackes and shales and derived drifts	LP Linhope		DO Dod		ER Ettrick	HR Hardlee
Rowanhill	Clayey till derived from Silurian and Ordovician greywackes and shales / Till derived from Carboniferous shales, sandstones, cementstones and coals / Till derived from the above named rocks with partially sorted surface layers				WN Winton, ME Macmerry		
Tynehead	Drifts derived from Carboniferous sandstones and Ordovician greywackes		TY Tynehead			FB Frostineb	CM Cakemuir
Yarrow	Gravels derived from greywackes of Silurian and/or Ordovician age	YW Yarrow					

Other soils:

BPT Blanket Peat	PT Basin and Valley Peat	OC Restored Open cast	AL Alluvium

(Adapted from Soil Survey map of Edinburgh and Peebles, Soil Survey of Scotland. H.M.S.O.)

Using the maps

There are hazards in using the maps, mostly arising from soil variation.

Mapped boundaries between soil series may, in the field, be very sharp, with a transition over a few metres with no intergrade soils. This can occur where there is an abrupt parent material change as where Bunter Sandstone and Magnesian Limestone formations meet. However, boundaries are often much more diffuse when they depend on microclimate and topography changes and all 'lines on the map' should be interpreted with care.

A more serious hazard concerns the 'purity' of soil series, which varies considerably. Because of the scale of mapping, it is impossible to indicate small areas of one series occurring within another. Some defined series as mapped on a scale of 1 : 63 000 or 1 : 50 000 are relatively pure, as much as 70 per cent of the area being closely similar to the standard profile description. Examples of this are the Sherborne series, a Brown Calcareous soil derived from Inferior Oolite and, slightly less 'pure', the Denchworth series, a Surface-Water Gley derived from Cretaceous and Jurassic clays (60 per cent pure). Other series, in some glaciated areas, are likely to be less than 50 per cent pure, and can include unmapped enclaves of several other series. At first sight such mapping of soil series would seem to be of little value especially if, as in some earlier memoirs no mention is made of purity. In

practice, however, the Sherborne 'impurities' are very similar in subsoil texture and structure to true Sherborne so that they might well behave similarly under agriculture. Also some series' 'impurities', depending on the mapper, simply reflect a parent material of slightly different origin.

It is sufficient to conclude that the soil series should *not* be regarded as an area of blandly uniform soil.

The ultimate horror for the soil surveyor must be to resort to mapping an area as a 'complex'. Tidy-minded people may regard this as a defect whereas it is simply a statement that the soil pattern is so complicated or occurs in such small units as to be unmappable on the scale in use. Such complexes can usually be described very satisfactorily in words. They usually present major farming problems such as variable drainage patterns and consequent uneven ripening of crops, as for example in the Whitsome Complex in Berwickshire based on a very intricate pattern of glacial till, Old Red Sandstone and Silurian greywackes.

Derived maps

The basic soil maps and memoirs need skill and experience in interpretation and are not easily used by the non-specialist. In recent years simpler maps have been derived from the basic maps for specific purposes.

Land-use capability maps

These are the most widely produced 'derived' maps. They originated in the USA where, before the Second World War, road engineers, bankers, tax men and urban planners recognized the importance of the maps in decisions on land-use and land values. It was not until later that they were extensively used in agriculture.

Land-use capability maps are produced in collaboration by soil surveyors, soil advisers and agronomists. Soil series with similar properties for agriculture are bulked together and the soil map is further modified by consideration of gradient, climate, ease of management, stock and crop performance. Thus, much of the complexity of the basic soil map is eliminated and parent material, while obviously still retaining a strong influence, is not a dominant criterion.

Figure 4.3 shows the land-use capability map covering the same area as the basic soil map in Fig. 4.2

There are seven classes becoming progressively poorer for agriculture and the classification is intended mainly for farming but can also be used by town planners, bankers, recreation planners, and others interested in land-use. It is important to stress that these are *capability* maps, the assumption being that management aspects of husbandry such as lime and fertilizer application and weed control will be well handled to achieve the potential of the land. This assumption is expressed as 'a moderately high level of

management'. Land is divided into seven classes according to the severity of limitations to its use.

- Classes 1–4 cover land suitable for arable crops, with limitations increasing from 1 to 4.
- Classes 5 and 6 are not suitable for arable crops but are useful for grazing and forestry.
- Class 7 is unsuitable for arable crops, grazing or forestry.

Each class is split into sub-classes according to five interrelated limiting factors, one or more of which might be operating in a particular parcel of land. They are soil-rooting zone (s), wetness (w), climate (c), gradient (g) and erosion (e). The maps are produced in colour code but also, printed in each unit, will be one or more of these symbols, so that a soil may be coded on the map as 3 sc indicating Class 3 soil with limitations in soil and climate. This classification is based very largely on *physical* limitations to cropping.

Class 1   Land with either very minor or no physical limitations to use. This land permits the use of a very wide range of crops which will produce good yields with moderate fertilizer rates. It is either well supplied with plant nutrients or responds well to fertilizers. Sites are level or gently sloping and climate favourable. The soil is well drained and deep (more than 75 cm). Textures are usually loam, sandy loam or silty loam. Root penetration is

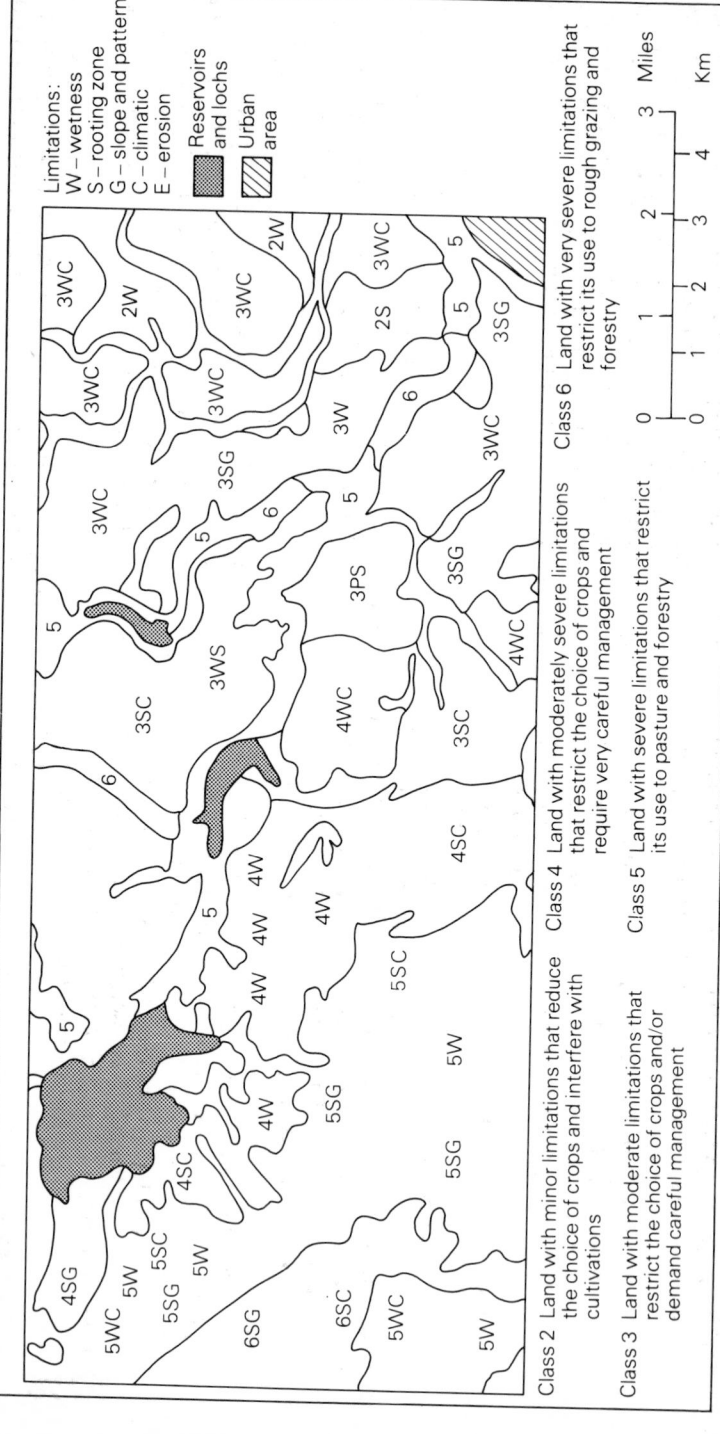

Limitations:
W – wetness
S – rooting zone
G – slope and pattern
C – climatic
E – erosion

▨ Reservoirs and lochs

▧ Urban area

Class 2 Land with minor limitations that reduce the choice of crops and interfere with cultivations

Class 3 Land with moderate limitations that restrict the choice of crops and/or demand careful management

Class 4 Land with moderately severe limitations that restrict the choice of crops and require very careful management

Class 5 Land with severe limitations that restrict its use to pasture and forestry

Class 6 Land with very severe limitations that restrict its use to rough grazing and forestry

Figure 4.3 A land-use capability map – based on Fig. 4.2

excellent and reserves of moisture are good. It is not surprising that in many parts of Britain no Class 1 soil appears on the maps.

Class 2 Land with minor limitations that reduce the choice of crops and interfere with cultivations. Class 2 is more widespread than Class 1 and, in many areas, represents the best land available. A wide range of crops can be grown but there may be harvesting difficulties for root crops. Limitations may include, singly or in combination:

w – moderate or imperfect drainage;
s – rooting depth slightly less than ideal but still 50 cm or more, structure and texture slightly unfavourable;
g – moderate slopes, not more than 7°;
e – slight erosion by water or wind;
c – slightly unfavourable climate, altitude usually below 230 m.

Class 3 Land with moderate limitations that restrict the choice of crops and demand careful management. Class 3 covers very large areas in the north and west of the British Isles and there might well be a case for dividing it as there is a large range of cropping potential within it. In this class there are definite limitations to the timing of cultivations and to the choice of crops. Arable land is mainly in cereal and forage crops and grassland is needed in the rotation.

There are usually at least two limiting factors from among:

w – imperfect or poor drainage;
s – restriction in rooting depth but still 25 cm or more, unfavourable structure and texture;
g – strongly sloping land, not more than 11°;
e – slight erosion;
c – moderately severe climate, more than 1000 mm annual rainfall, usually below 380 m altitude.

Class 4 Land with moderately severe limitations that restrict the choice of crops and require very careful management. The main crop is grass with occasional breaks into cereals or forage crops but certainly choice of crops and yield are very restricted. Defects are:

w – poor drainage difficult to remedy, occasional damaging floods;
s – shallow or very stony soil but capable of being ploughed;
g – moderately steep gradient (11–15°);
e – slight erosion;
c – moderately severe climate, rainfall more than 1250 mm, altitude usually below 450 m.

Class 5 Land with severe limitations that restrict its use to pasture, forestry and recreation. This class occupies large areas in the north and west to the British Isles in which grazing, agriculture, forestry and recreation either compete or come to a reasonable compromise on land-use. Arable cropping is virtually impossible. Whereas the limiting factors in Classes 1–4 are usually economically correctable this is not so in Class 5. There are usually several defects:

w – poor or very poor drainage, frequent damaging floods;

s – soil too stony or shallow to plough;

g – very steep slopes 15–20°;

e – severe risk of erosion;

c – severe climate, altitude 450–530 m.

Class 6   Land with very severe limitations that restrict use to rough grazing, forestry and recreation. The only aspect of agriculture on this land is rough, usually very steep grazing. Soil is often peaty, slopes steep and erosion severe.

Class 7   Land with extremely severe limitations that cannot be rectified. This is usually either peaty or very rocky with very steep gradients and extremely severe climate. Sites lie above 600 m. Snow cover for long periods and the very short growing season prevent even forestry.

Uses of land-use capability maps

Now that general land-use capability maps are becoming widely available, they are being used increasingly by planners, advisers and farmers. They are useful in assessing land values for farm purchases and for deciding between alternative land-uses on a large scale, for example, grazing or forestry.

There are also applications in farm planning. Root crops, for example, may be grown successfully on Class 1 and Class 2 land but the limitations of Class 3 affect sowing, growing and harvesting and root crops should be grown only if the need to produce the crop outweighs the problems.

The classification also leads to a better appreciation of soil management needs. Obviously good management is required on all classes of land but Classes 3 and 4 demand very sensitive and patient handling for good results. Soils in these two classes cover very large areas in northern England, southern Scotland, northern Ireland and parts of Wales.

The classification used in the United Kingdom is strongly orientated towards cropping and it pays scant attention to the needs of the stock farmer. No doubt this drawback will be rectified. Meantime, decisions on which land to allocate to grazing and on stocking intensity can only be helped indirectly by the classification.

Special-purpose maps

Many kinds of special-purpose maps may be produced

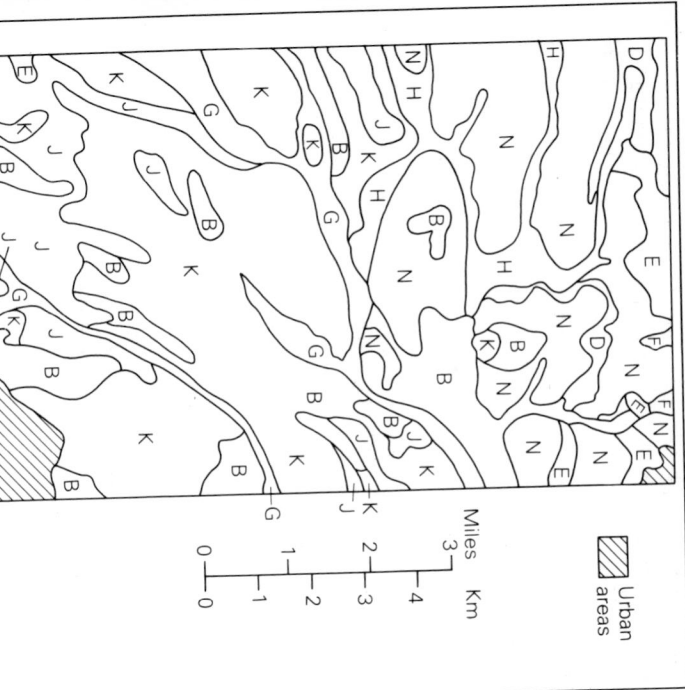

Group symbol	Type of system	Tile interval	Depth
B	Pipe drains without permeable fill	20–40 m	>90 m
D	Random	20–30 m	75–90 cm
E	Pipe drains with permeable fill. Use of permeable fill related to the incidence of clayey subsoils	30–50 m	90–120 cm
F	Use of permeable fill related to the incidence of clayey subsoils		90–120 cm
G	Spacing and depth related to outfall conditions: land often too poor for arable use but where good outfall conditions exist and there is no flooding risk mole-cum-tile schemes are feasible		
H	Spacing and depth related to outfall conditions: use of permeable fill related to incidence of peaty soils	20–40 m	75–90 cm
J	Moling over piped mains with permeable fill	20–40 m	>90 cm
K			
N	Drainage not generally required for arable land use		

Miles
0   1   2   3
Km
0   1   2   3   4

Urban areas

Figure 4.4 . A special-purpose map predicting underdrainage treatments for arable land-use (Adapted from part of map 'Predicted underdrainage treatment for arable land use' Soils and Field Drainage, Technical Monograph **7**, Soil Survey of England and Wales. H.M.S.O. 1975)

Table 4.2 Examples of special-purpose maps derived from basic soil maps

Purpose of map	Examples
Forecasting the risk of problems	Erosion Capping Drought Waterlogging Cultivation pans Poaching Copper deficiency Wind-blow (forest trees)
Assessing the need for treatment	Irrigation Subsoiling Drainage (type and spacing)
Assessing suitability or capability for general or specific purposes	Land-use capability (general) Grazing capacity Proportion of arable crops in rotation Direct drilling Minimal cultivations Specific crops: carrots, maize, soya bean, apples

The authors add a very wise rider to their map urging the drainage designer that 'each site will require thorough investigation, that account must be taken of economic circumstances and site factors including climate, outfall conditions, slope, permeability and existing underdrainage'.

These provisos are absolutely necessary for any special purpose map, which is best used to create awareness, in the mind of the user, of problems and risks.

Maps at present available can be grouped into three types – concerned with risks, needs and suitability. Table 4.2 indicates some of the maps which are available.

from the basic soil map, using also the local knowledge of advisers and analytical data from the memoirs. Figure 4.4 is part of a derived map showing 'Predicted underdrainage treatments for arable land-use' in the Abingdon area. The key to the map gives predictions for drain spacing and depth and the need for 'permeable fill' (usually gravel).

Soil organic matter

The type and quantity of organic matter in soil have very large influences on the structure of the soil and hence the ability of fine feeding roots to penetrate in search of nutrients, the capacity of the soil to hold available water for plants, the prevention of excessive leaching and the ability of the soil to retain nutrients in forms available to the plant. These points alone indicate that it is essential to maintain or, where possible, increase the quantity of organic matter in mineral soil.

There is a very vocal school of thought which propagates the idea of 'organic farming'. It is a pity that they have, in some cases, taken their ideas to the extreme by opposing the use of 'chemical fertilizers' for they have indeed grasped on to a great truth about soil and the critical role of organic matter in its fertility, but there is nothing incompatible about maintaining organic matter levels and the use of fertilizers. The simple fact is that, although organic farming may be self-supporting in certain areas, the crop yields required to produce sufficient food for the world could not possibly be obtained under this system without the use of fertilizers.

Raw material

The raw material of soil organic matter consists of plant and animal remains and animal excreta. Most soil organic

Table 5.1 Composition of plant material

	% of plant dry matter
Simple sugars, pectins	1– 6
Starch, hemicelluloses, polysaccharides	12–30
Cellulose	20–50
Nitrogen compounds:	2–18
proteins, amino-acids	
Fats, oils, waxes, tannins	1– 7
Lignins	10–30
Mineral matter	1– 5

matter is derived directly from plant residues which are eventually decomposed to form humus. The composition of the raw material varies considerably according to whether it is young or old, leaf or stem, or in the case of trees, leaf, bark or trunk. It will also vary from species to species. Leguminous plants such as peas, beans or clover will produce organic matter much richer in nitrogenous compounds than that from non-legumes.

Table 5.1 shows the groups of substances that make up plant material and the wide variation in the amounts present. The substances in the top three groups are composed of carbon, hydrogen and oxygen. They are relatively simple compounds and are used by soil organisms to supply the carbon they need both for energy and for building body tissue.

The proteins and amino-acids contain, in addition to carbon, hydrogen and oxygen, nitrogen and sulphur which are also needed by the soil organisms to produce their own body protein.

The substances included in the lower part of Table 5.1 are complex organic compounds which are resistant to microbial decomposition.

Decomposition of organic matter

As soon as raw organic matter is deposited on, or in, the soil, decomposition begins. It is a very complex process, almost entirely brought about by the activities of soil organisms.

The early stages of decomposition are dominated by macro-organisms such as beetles, mites and earthworms which eat and excrete the organic matter. As an example of this, the number of earthworms in a rich permanent pasture can be as high as 2 million per hectare and such a population can ingest and excrete 100 t/ha of soil every year. The organic matter in the excreta of the macro-organisms has been partly decomposed and is in smaller pieces.

During this early stage of decomposition, some micro-organism activity also occurs. Fungi and bacteria act upon the organic material throughout decomposition. Their numbers are vast, being sometimes several million per gram of soil. The actual weight of the living micro-organisms in soil is about 2 per cent of the total weight of organic matter so that a soil with 5 per cent of organic matter will contain about 2.5 t/ha of active bacteria and fungi. For the most part the bacteria feed upon the more easily decomposable plant material. There is a rapid boost in numbers of bacteria whenever fresh organic matter is deposited in the soil. When the sugars, starches and cellulose have been exhausted the population declines again, quite rapidly. Some of the fungi also feed on more resistant plant material, even tackling highly resistant materials such as lignin.

There is, therefore, a period of rapid decomposition during the first few months, during which most of the sugars, starches, hemicelluloses and some cellulose are decomposed. Much of the carbon they contain is converted to carbon dioxide and lost into the atmosphere. After this a slower phase of decomposition takes place during which the remainder of the cellulose and some of the more resistant materials are decomposed. Much of this material remains in the soil, although changed, and some of it has been shown, by carbon-14 dating, to persist in the soil for thousands of years. The persistent part of soil organic matter is known as humus, a black or dark brown mixture of organic compounds some of which are very complex. The decomposition processes can proceed efficiently only if conditions allow vigorous soil organism activity. This requires adequate but not excessive water availability, ample oxygen in the vicinity of the fresh organic matter and an alkaline or mildly acid pH (5.5–7.5). Thus either drought or waterlogging will restrict humification. Another all-important factor is the amount of nitrogen, in the form of proteins, amino-acids or simple inorganic nitrates and ammonium compounds, that is available for the micro-organisms.

In undisturbed soils where no fertilizer nitrogen is added the main source of nitrogen for micro-organisms is decaying plant and animal material. The nitrogen requirement for the micro-organisms to live and multiply is high, especially during the population explosion that follows addition of fresh organic matter to the soil.

The importance of carbon/nitrogen ratios

The ratio of carbon to nitrogen in microbial cells is approximately 6. If the amount of carbon in the decomposing organic matter greatly exceeds that of nitrogen the organisms cannot maintain the ratio of 6 for their body requirements, the population is restricted and humification is retarded or stopped. In other words, the higher the carbon/nitrogen ratio of the organic matter, the slower and more incomplete is the humification.

Some typical carbon/nitrogen ratios of plant materials,

Table 5.2 Carbon/nitrogen ratios of plant materials, manures and soils

Type of material	Range of carbon/nitrogen ratios
Plant:	
Young clover leaves	12:1–18:1
Young cereal and grass leaves	15–20
Barley straw	40–80
Cereal roots and stubble	40–60
Legume stubble	20–25
Leaves of broadleaved trees	25–60
Leaves of coniferous trees	80–120
Sawdust from coniferous trees	300–400
Manure:	
Fresh farmyard manure	30–40
Well-rotted farmyard manure	15–25
Slurry	15–18
Soil	
Brown Earth (mull humus)	10–12
Podsol (mor humus)	25–35
Podsol (litter layer)	50–120
Arable soil	10–12

manures and soils are given in Table 5.2

The lowest carbon/nitrogen ratios occur in the leaves and root nodules of young leguminous plants. They are readily decomposable. Materials such as grass or cereal leaves will decompose less readily. Straw is yet more difficult to decompose; the leaves and sawdust from coniferous trees very difficult indeed.

The only way in which soil organic matter with a high carbon/nitrogen ratio can be decomposed is by the use of nitrogen from elsewhere in the soil system. This might come from indigenous soil nitrogen, from fertilizers or manures. When this happens the micro-organisms are competing for available nitrogen with plants. This can result in nitrogen deficiency in crops grown within a few months, even a year, after the addition of high carbon/nitrogen ratio material such as straw. Extra fertilizer nitrogen is needed to correct the deficiency.

Types of humus

Mull Mull humus is characteristic of Brown Earths and Rendzinas. It is dark brown or black, intimately mixed with mineral matter and forms stable clay–humus complexes with high cation exchange capacity. Because of the rapid humification, encouraged by ample but not excess water, good aeration and neutral or mildly acid pH, there is little or no build-up of 'litter' on the soil surface.

Mor and moder These humus types are formed where acidity restricts humification and incorporation of organic matter and are found in Podsols or Brown Earth–Podsol intergrade soils. They are characterized by the build-up of purely organic horizons on the surface of the mineral soil caused by poor humification and restricted organic matter incorporation resulting from acidity.

Three types of pure organic horizons can be distinguished:

L – the litter layer, undecomposed plant debris.

F – sometimes called the fermentation layer. It is darker in colour than the L-layer, and is usually medium or dark brown. Some decomposition has occurred but plant remains can be clearly seen. In the field the F-layer can be 'peeled off' from the underlying H-layer by hand.

H – the humified layer, generally black. It contains few identifiable plant remains and can be compact and greasy when wet and hard when dry.

Mor humus has well developed L- and F-layers but the H-layer may be thin or absent.

Moder humus is transitional between mull and mor and, as such, occurs on intergrade soils between Podsol and Brown Earth. It is distinguished from mor by the presence of a well-developed H-layer usually associated with only thin F- and L-layers.

Figure 5.1 shows the transition of mull–moder–mor as the pH declines in a group of well-drained undisturbed soils. The deeply incorporated mull humus gives way first to less deep incorporation with a thin but discrete H-layer. As conditions become more acidic the H-layer thickens (moder humus) and the amount of incorporation declines.

Then, as neither decomposing nor mixing organisms can tolerate the increasing acidity, a new discrete horizon (F-layer) of partially decomposed organic matter builds up and replaces the H-layer (mor humus). Finally in extremely acid conditions an L-layer of completely undecomposed material builds up. In some cold, very acid coniferous forests in Scandinavia the L-layer can be 20 cm in depth and the soil is virtually sterile.

Peat   Peat is formed where waterlogging restricts humification and organic matter incorporation. The development and characteristics of various types of peat have been described in Chapter 3.

Cultivated soils   The type of humus which develops in a well-drained, well-limed, well-fertilized arable soil is similar to mull humus formed under undisturbed conditions. It will have a carbon/nitrogen ratio of about 10 and will be intimately mixed with soil mineral matter. Even cultivated soils previously covered by mor, moder or peat will tend to be converted to mull-like humus after reclamation for agriculture.

The stability of this humus type, as with any other type, depends on the maintenance of the conditions under which it was formed and came to equilibrium in the soil. Obviously great changes are caused by some agricultural processes.

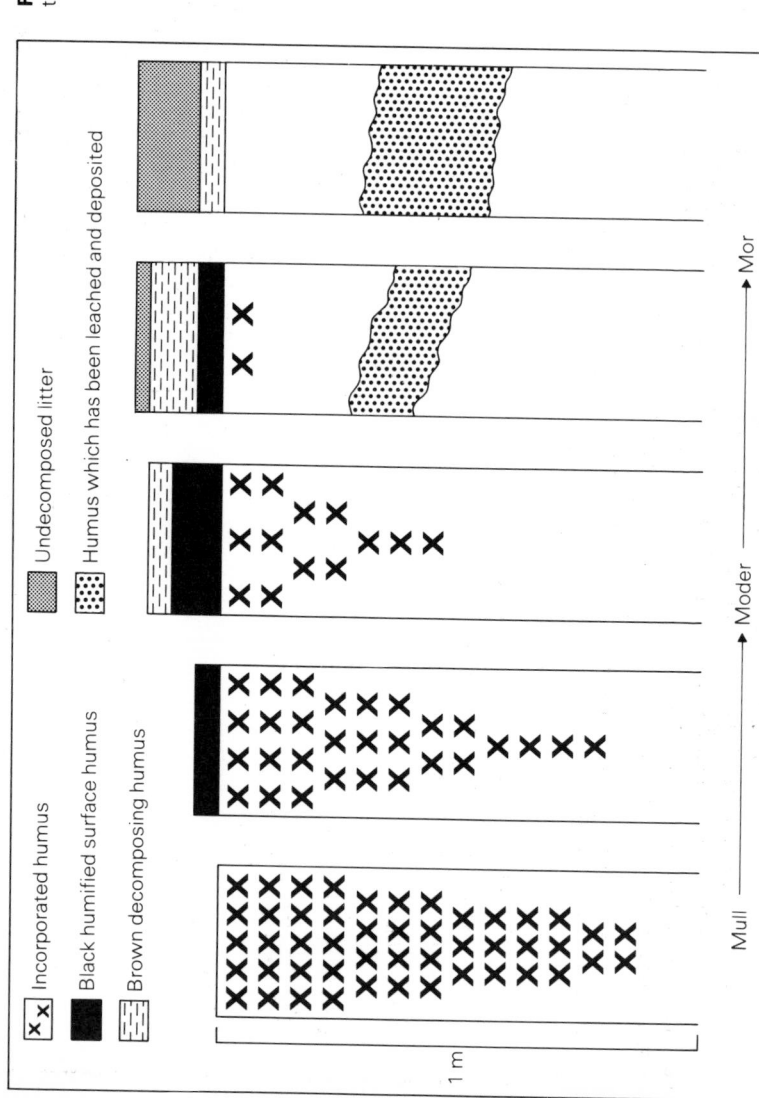

Figure 5.1 Humus types – the transition from mull to mor

Losses of organic matter from arable soils

Liming by raising soil pH, drainage by increasing aeration, fertilizer application by increasing available nutrient levels (especially nitrogen) and cultivations which can either increase or decrease aeration, all radically change soil conditions. Only permanent pasture and arable soils with minimal cultivations escape the most severe changes.

The net result when a soil is reclaimed for agriculture is a very sharp and fairly rapid decline in the quantity of organic matter the soil contains. The reason for this decline, which has been recorded all over the world, is greatly increased oxidation of organic matter, aided by microorganism activity. The result is the loss of large amounts of carbon dioxide into the atmosphere as the organic matter is destroyed.

Figure 5.2 gives examples of the rate of loss and new 'equilibrium' levels of organic matter following reclamation. Both the initial rate of loss and the final organic matter content of the soil vary widely according to conditions. The rate of loss of organic matter is greatly influenced by soil temperature, aeration and availability of water. Soil acidity will reduce the rate of loss but, in agricultural soil between pH 5.5 and 7.5, has little effect. Most rapid losses occur when adequate available water is constantly present with good aeration and high soil temperatures. Typical examples of this are the rapid losses of organic matter from the fen soil of East Anglia and

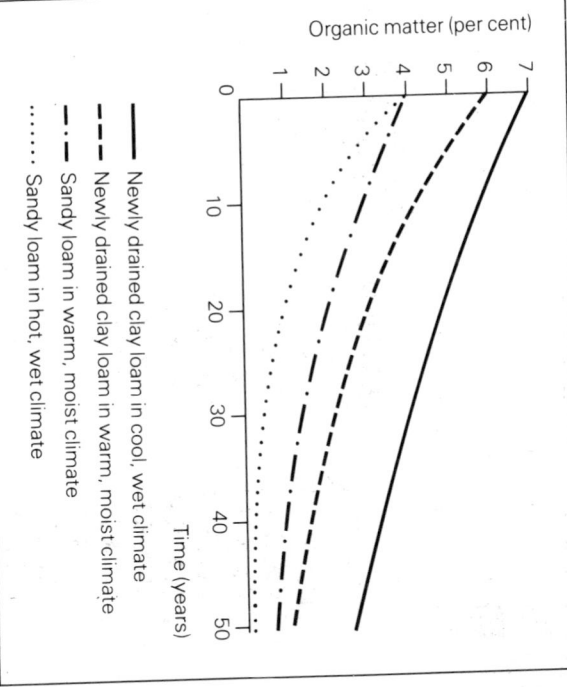

Organic matter (per cent)

——— Newly drained clay loam in cool, wet climate

– – – Newly drained clay loam in warm, moist climate

–·–·– Sandy loam in warm, moist climate

········· Sandy loam in hot, wet climate

Time (years)

Figure 5.2 Loss of organic matter from soils following reclamation

similar peat soil south of Lake Erie in USA where water-table control ensures aeration and availability of water, the pH is kept high by incoming water from calcareous formations, and intensive cultivations and fertilizer application have been practised.

The organic matter lost in these circumstances is mainly material with a fairly high carbon/nitrogen ratio, the decomposition of which has been restricted in the undisturbed state. As this material is decomposed the rate of loss will slow down and eventually a new equilibrium organic matter content will be established depending on the type of arable agriculture. Total loss of organic matter is unlikely to occur because of the presence of new crop residues and, more important, the existence of 'protected' organic matter. There is strong evidence that much of this protected material is hundreds or thousands of years old. It is known to be strongly linked with clay and some of it is probably situated in pores too fine for micro-organisms to enter. Protected organic matter will therefore be more likely to exist in soil with relatively high clay content. The organic matter in sandy or silty soil will be more vulnerable. In fact organic matter contents are generally lower in light-textured than in adjacent heavy-textured soil.

If organic matter losses from cultivated soils lead to critically low humified organic matter contents, varying from 1 to 3 per cent according to local conditions, serious soil management problems can occur in soils of all texture groups. In heavy soils, low in organic matter, excessive plasticity and cohesion can lead to cultivation problems such as clod formation, plough pans and limitations to the period available for cultivations without damage to soil

structure. Fine sandy soils containing little organic matter can be subject to wind erosion. Silty soils low in organic matter will develop 'capping' after heavy rain with adverse effects on seedling emergence.

Beneficial effects of soil organic matter

Organic matter, particularly if well humified, has profound effects on several soil properties which govern overall fertility and the potential for crop production. In many years the beneficial effects show up simply in ease of cultivations permitting early sowing, good seedling emergence, rapid early growth and early harvest. If there are extremes of drought or wetness the modifying effects of organic matter are much more marked.

Maintenance of soil structure

Humus has a critical role in the formation and maintenance of good soil structure (crumb, granular, sub-angular blocky – see Chapter 6). This helps to maintain a high available water capacity and to reduce the plasticity and cohesion of heavy soils.

Retention of available nutrients

Because of its colloidal properties and high cation

Table 5.3 Cation exchange capacity of organic matter and clays

Material	Cation exchange capacity (milliequivalents/100 g soil)
Organic matter	
Well-decomposed humus (mull, arable soils)	200–300
Humus of poorly drained or acid soils (mor, peat)	30–100
Clays	
Montmorillonite	80–100
Illite (hydrous mica)	15–40
Kaolinite	3–15
Hydrous oxide	3–5

exchange capacity, well humified organic matter increases the ability of the soil to retain available calcium, magnesium, potassium, ammonium and several important trace elements against leaching. In this respect, the organic matter is two or three times as effective as the most retentive type of clay, montmorillonite, and six to ten times as effective as the illitic clays which are more commonly present in the soils of the British Isles (Table 5.3).

Less well-humified types of organic matter such as mor, moder and peat have relatively low cation exchange capacities, approximately equivalent to that of montmorillonite.

Thus the quality and quantity of humus have profound influences on nutrient retention. This is illustrated in Fig. 5.3 which compares the cation exchange capacity of soils of different texture containing various amounts of 'good' or 'poor' humus. The influence of organic matter is greatest in light-textured soils, the cation exchange capacity of the sandy loam (Fig. 5.3) being more than doubled by increasing the organic matter content from 1 to 5 per cent. In practice, organic matter accounts for 10 to 80 per cent of the cation exchange capacity of soils in the British Isles, most of the higher values occurring in the cooler wetter regions.

High water-holding capacity

The combination of the sponge-like properties of humus and its effects on soil structure gives rise to increases in the available water capacity of mineral soils. Well-humified organic matter, as in most agricultural soils, can retain two to four times its own dry weight of water, about half of which will be available to the plant. The presence of 5 per cent of such humus will, therefore, increase the available water capacity of a sandy loam by more than 50 per cent and that of a clay loam by about 30 per cent above the levels in comparable organic-matter-free soils. This effect will be particularly important in times of drought.

Some poorly decomposed peats derived from sphagnum

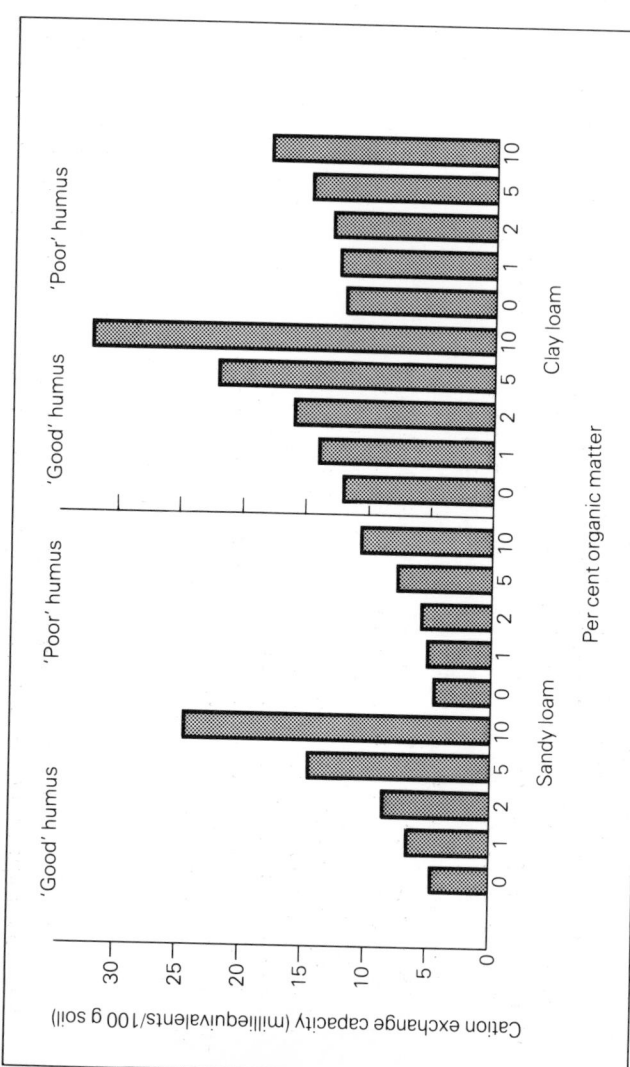

Figure 5.3 The cation exchange capacity of soils of different texture

moss retain 10 to 12 times their own dry weight of water. In the field this is a disadvantage and can lead to serious problems of poor aeration. Such peats are, however, valuable to the gardener who mixes them with mineral soil to increase water retention.

Low plasticity and cohesion

Unlike clay colloids, organic colloids have low plasticity and cohesion. Particularly in clay, clay loam and silty clay soils, the maintenance of high humus contents will help to alleviate the cohesive and plastic qualities of clay which contribute to poor structure and give rise to problems such as clod formation and pan formation during cultivations.

Gradual release of available nutrients

During the early stages of decomposition of fresh organic matter, especially of a low carbon/nitrogen ratio, essential nutrients are released by a process known as mineralization. Both major and trace elements, originally present in the plant or animal tissue in complex forms, are released as simple cations or anions. The main elements involved are nitrogen, phosphorus and sulphur. It is important to note that this release of nutrients occurs through the action of soil micro-organisms and will *not* occur if their action is restricted by high carbon/nitrogen ratios, by soil acidity or poor aeration.

Darkening of colour

Humus types, except some very acid sedge peats which are pale yellow, are brown or black. They impart these colours to soil. An outstanding example of this is the Chernozem or Black Earth soil developed in the short-grass prairies of North America and the Eurasian steppe, with its rich black humus. More commonly organic matter is responsible for the drab colours of many types of soil. In Britain this effect may be seen on the outskirts of many villages where the practice of carting human excreta as 'night soil' went on for many centuries, greatly modifying the soil and darkening its colour.

This humus-darkened soil is able to absorb more heat from the sun. Very early research indicated that creating a dark soil surface by applying soot could increase the soil temperature by 1–3 °C at a depth of 15 cm. Such temperature differences are important in earliness and speed of growth of crops before the cover becomes dense. This has been recognized in horticulture by the use of black polythene soil cover for the production of high-value crops.

Organic matter content

The organic matter content of soils varies from less than 1 per cent, in young soils and those exhausted by over-intensive cultivation and cropping in hot dry climatic conditions, to 95 per cent in some deep peats. Soils with more than 25 per cent may be regarded as 'organic' and those containing less than 25 per cent as 'mineral'. The great majority of cultivated mineral soils contain less than 10 per cent of organic matter in the surface horizons.

Table 5.4 shows the average and range of organic matter contents of groups of soils in south-east Scotland under different management systems. Although the organic matter contents of the different groups overlap the average values show the strong influence of grassland in maintaining high organic matter.

Similar ranges and average values to those shown in Table 5.4 would be found for large areas of northern England, Wales and Ireland. Much lower values are found in south-east England, less than 2 per cent being common for intensive arable land. The organic matter contents of the soils on the ley/arable experiments at the ADAS Experimental Husbandry Farms in England and Wales vary, for the arable rotations, between 5.6 per cent at Trawscoed in Wales and only 1.5 per cent on a sandy soil in the east Midlands.

Table 5.5 gives a guide to the organic matter status of

Table 5.4 The organic matter content of mineral soils in south-east Scotland

Type of agriculture	% organic matter	
	Average	Range
Arable	3.8	2.0–6.2
Arable with 25% grass	4.4	1.9–7.4
Arable with 50% grass	5.0	3.6–6.8
Permanent pasture	8.5	7.9–9.5

Table 5.5 Organic matter status of mineral surface soils

Status	% organic matter
Very low	Less than 1.0
Low	1.0–1.7
Moderate – low	1.8–3.0
Moderate	3.1–5.0
High	5.1–8.0
Very high	More than 8.0

mineral surface soils. Subsoils usually contain much less organic matter than surface soils. Soils in the low or very low categories may be expected to present serious cultivation and erosion problems in some seasons.

Mineral soils with 'very high' organic matter contents are usually under permanent pasture or are cultivated Peaty Gleys.

Organic matter maintenance

The manifold benefits of organic matter to soil fertility indicate the wisdom of maintaining or building up the content of well-decomposed humus. It is much more difficult to do this than it is to destroy organic matter but there are many simple procedures that can be adopted, within economic systems of agriculture which will help to maintain soil organic matter. These are described in Chapter 10.

Soil texture and structure

6

The physical condition of soils is dominated by two closely related properties – texture and structure. The two are often confused and a clear understanding of the difference between them is essential.

Texture is determined by the sizes and proportions of the small individual mineral particles in the soil – sand, silt and clay. Structure is built up by the aggregation of textural particles into larger units through the influence of agents such as organic matter, earthworms and other soil fauna, wetting and drying, freezing and thawing and the pressure of roots.

Texture

The term texture probably arose from the feel of moist soil when rubbed between the fingers and thumb. Each texture type has a characteristic feel. Silty soils feel smooth and silky, whereas sandy soils feel gritty and clay soils feel sticky and plastic when wet.

To help classify textures, soil scientists have defined the borderlines between the various groups of 'fine earth' particles which make up soil. Fine earth includes only those particles less than 2 mm in 'diameter'. Particles larger than this – gravel, pebbles and stones – are not included in texture assessments and separate account must be taken of their effects on soil properties.

Table 6.1 The size and properties of soil particles (international classification)

Particle	Size (mm)	Properties	Material with particles of comparable size
Gravel	More than 2	Particles plainly visible	Granular fertilizers–hazelnuts
Coarse sand	0.2–2	Particles plainly visible Very gritty	Coarsest sandpaper, builders' sand
Fine sand	0.02–0.2	Particles distinguishable without lens Slightly gritty	Finest sandpaper, egg-timer sand, castor sugar
Silt	0.002–0.02	Individual particles identifiable using hand lens	Fine white flour, icing sugar
Clay	Less than 0.002	Smooth silky feel, wet or dry Individual particles identifiable only by microscope Sticky, plastic and cohesive when moist, drying to a hard mass	Plasticine, putty

Table 6.1 shows the internationally accepted borderlines between size groups. There are other systems of classification which use slightly different borderlines, the only important difference being that the United States Department of Agriculture use 0.05 mm as the dividing line between sand and silt.

'Sand' as we usually imagine it on beaches or in sand quarries is mainly coarse sand and may contain some gravel. Particles of fine sand, as defined, are very much smaller and occur widely in dunes and other wind-borne deposits. Both coarse and fine sand can be seen as individual particles by the naked eye. Silt particles cannot. Clay particles are very small indeed and are colloidal in nature.

Soil containing large proportions of sand and gravel has large pores between the particles through which water can drain freely. As the proportion of fine sand or silt increases the total volume of pores increases but the size of individual pores is smaller and the soil is able to retain more water for plant growth. Soils containing large amounts of clay retain even more water in the large volume of very small pores.

The importance of the clay fraction

Unlike sand and silt, much of the clay fraction has colloidal properties. The particles are very small and, if dispersed, the smallest of them can remain suspended in water for years. Clay has electrical charges on its surfaces, mostly negative, which cause it to attract positively charged ions *cations*. This process is known as adsorption and the capacity of the clay to retain cations is called the cation exchange capacity. Cationic plant nutrients such as potassium (K^+), magnesium (Mg^{2+}), calcium (Ca^{2+}) and ammonium (NH_4^+) are held by the clay sufficiently strongly to prevent them from being easily leached. Plant roots can, however, remove and absorb them so that they are 'available' to the plant.

Because of its colloidal properties clay retains water. As it attracts and absorbs water, the clay swells. If the water is removed by plant roots, the clay shrinks. This is important in determining the type of structure that is formed.

As clay becomes wet it becomes plastic, capable of moulding, and cohesive, the individual particles tending to unite into one mass. These properties lead to cultivation problems in soils containing a high proportion of clay.

Types of clay minerals

There are many types of clay minerals, depending on the rock minerals from which they have been formed and the weathering processes involved. Most clay minerals are alumino-silicates and are plate-like and crystalline in form.

The type of clay can considerably affect the ability of the soil to retain nutrients, and can affect swelling and shrinking, and plasticity and cohesion. There are three major types of clay colloid: montmorillonite, illite (or hydrous mica) and kaolinite.

Montmorillonite clays have high cation exchange capacity (Table 5.3) high plasticity and cohesion, high swelling capacity and high water retention. Kaolinite has low cation exchange capacity, is not strongly plastic or cohesive, does not swell and shrink greatly and has lower water retention. Illitic clays are intermediate in all these properties.

In the soils of the British Isles the dominant clay minerals are commonly intermediate in properties between montmorillonite and kaolinite. The modifying influence of organic matter on water and nutrient retention, plasticity and cohesion is considerable. This tends to mask the effects of any differences in clay type.

Classification

It is fairly easy to distinguish between 'light' (sandy) soils and 'heavy' (clay) soils but a large proportion of the British Isles is covered by soils of medium texture. If you ask a farmer who works on such soil, he will probably call it loam or 'a loamy soil'. Unfortunately the term 'loam' is used, depending on the general pattern of soils in the area to describe widely different texture types in different parts of the country. Here it will plough firm in the furrow and you will be able to walk across the tops of the drying ridges. Elsewhere it will fall loosely from the plough.

Fairly small differences in texture bring about important changes in management problems. It is, therefore, useful to be able to identify a range of texture types. This has been done by introducing terms such as sandy loam and silty clay.

The proportions of sand, silt and clay in soil can be estimated by various laboratory procedures and plotted on the texture triangle shown in Fig. 6.1(a). This shows the whole range of texture types. The three corners of the triangle represent 100 per cent clay (top), 100 per cent sand (bottom left) and 100 per cent silt (bottom right).

Results for two samples of soil are shown in Fig. 6.1(b). Soil X is a 'heavy' soil containing 35 per cent sand, 30 per cent silt and 35 per cent clay and is a clay loam. Soil Y is a 'light' soil, containing 70 per cent sand, 15 per cent clay and 15 per cent silt and is a sandy loam.

The majority of soils in the British Isles are grouped towards the bottom left-hand part and central parts of the triangle. Apart from some sedimentary formations in southern and eastern England, such as Kimmeridge,

Oxford and London clays, it is uncommon to find soils with more than 50 per cent clay. Silty textures are also relatively rare except in restricted areas such as the East Anglia fens and on the banks of estuaries such as the Humber.

Other points may be drawn from the texture triangle:

- Loam, as defined, is a very limited texture type containing 10–25 per cent clay, 30–50 per cent sand. 20–50 per cent silt.
- Soil containing as little as 40 per cent clay is classified as 'clay' and is indeed a very 'heavy' soil. Clay soil may contain as much as 45 per cent of sand and still be 'heavy'.

The real test of the usefulness of the texture triangle is whether the practical problems experienced by the farmer (difficulties in working, 'puffy' seedbeds, excessive compaction, plough pans) are well related to the defined texture types. This is broadly true for the soils of the British Isles. There are, however, some drawbacks to the use of the triangle. The most important is that it does not distinguish between coarse and fine sand. There is a 100-fold difference in size between the coarsest coarse sand and the finest fine sand. There are consequent effects on soil problems such as compaction and wind erosion, both much greater on soil containing high proportions of fine sand. The main texture type names can be adjusted to indicate the dominance of fine sand (fine sandy loam).

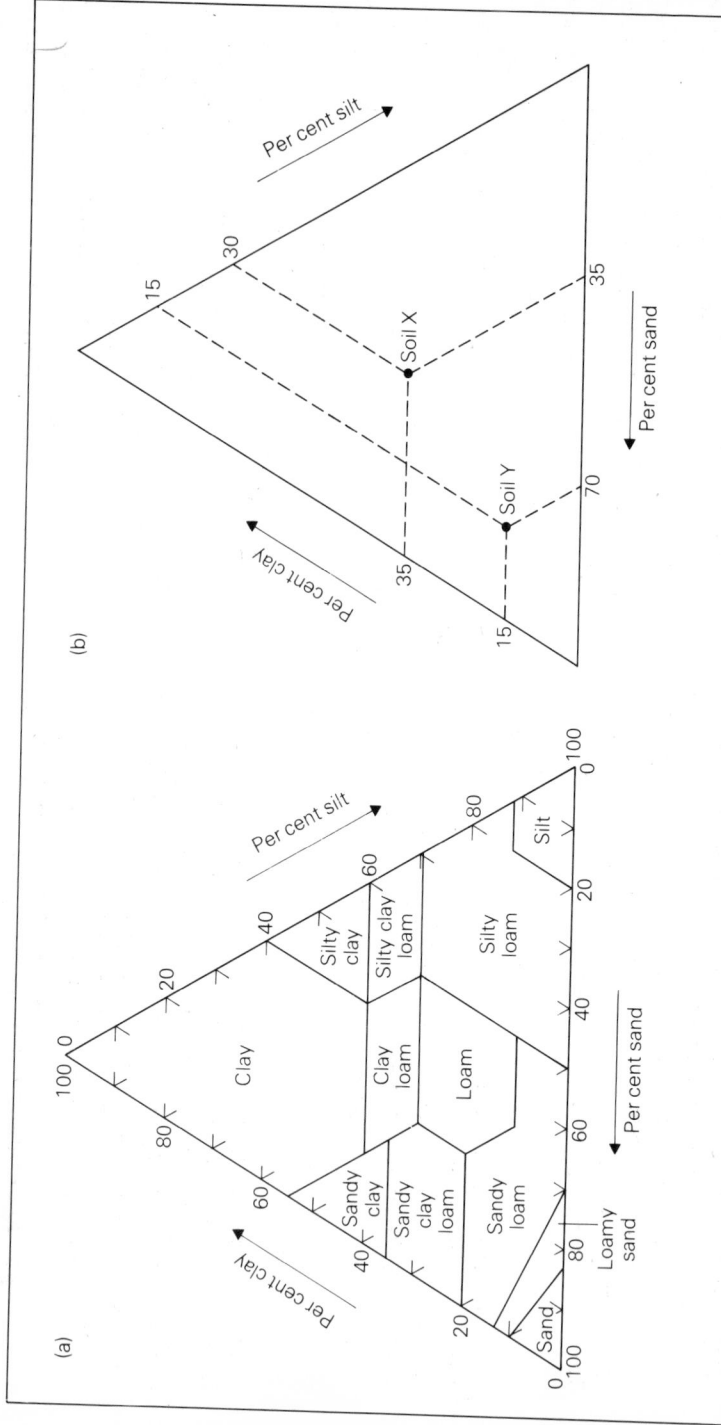

Figure 6.1 The 'texture triangle' – a method of classifying texture

Account must also be taken of particles greater than 2 mm in cross-section – stones and gravel, which can greatly modify the effects of texture.

The texture triangle also takes no account of the type and quantity of either organic matter or calcium carbonate, both of which can modify the effects of texture type. None the less there is a clear relationship between the position of a soil in the texture triangle and the problems of managing the soil.

Assessment of texture in the field

The laboratory and sampling procedures required to place a soil in the texture triangle are costly and tedious. With a little training and practice it is possible to distinguish most soil texture types in the field. The feel and appearance of sand, silt and clay are quite different.

Individual particles of coarse sand can be clearly seen. When rubbed together between the fingers in a moist condition the grains will grate against each other and feel very rough. Individual grains of fine sand are barely visible and the roughness and grating of a moist sample is less obvious, but still detectable. In silt, individual particles cannot be identified as they are too small. The moist sample of pure silt feels characteristically very smooth and 'silky'. It is slightly sticky when wet. Individual particles of clay cannot be distinguished as they are even smaller than

silt. Clay is very sticky when moist and can be rolled between the fingers into thin 'sausages'.

To assess the texture in the field, the soil should be examined, if possible, both dry and moist. For 'moist' examination, water should be added slowly to the dry soil until it darkens but excess water should be avoided. The moist soil is rubbed between the fingers and thumb for some time to break down the structural units and to assess the relative proportions of coarse sand, fine sand, silt and clay.

Characteristics of some texture types

Sand The soil is loose when dry and sand particles are plainly visible. It is not at all sticky when wet, but feels very gritty. It cannot be moulded or rolled into sausages. High proportions of fine sand can be identified by comparative lack of grittiness.

Sandy loam The sand fraction is plainly visible when dry, but the soil is not so loose. When moist the soil feels very gritty between the fingers. It can just be moulded, but does not retain any impressed shape easily. It does not stick to the fingers very much and cannot be rolled into sausages. High proportions of fine sand can be identified by comparative lack of 'grittiness'.

Loam The sand fraction is more difficult to distinguish when dry. The soil is slightly gritty when moist. It moulds readily and retains impressed shapes. It sticks to the fingers slightly. It will roll into sausages with difficulty but the sausages will break up if an attempt is made to roll them to diameters less than 2 mm.

Clay loam The soil is rather hard when dry. It is distinctly sticky when wet. The sand fraction can be identified only by pressing the moist soil hard between the fingers. It will roll into sausages with ease and the sausages will not break up so easily as loam as the diameter falls below 2 mm.

Clay The soil is very hard when dry and very sticky when moist. It is very difficult to detect the sand fraction by pressing the moist soil between the fingers. The soil can be rolled easily into fine sausages of less than 2 mm diameter.

Silt The dry soil has obviously very fine particles, but is not usually hard and feels smooth to the touch. The smooth silky feeling of silt is dominant when the soil is moist. It will form sausages by rolling between the fingers, but they will break up if any attempt is made to reduce the diameter below 2 mm.

Silt loam The soil is moderately hard when dry and has a similar appearance to clay loam, but is not very sticky when wet, has the characteristic silky feel of silt and will not mould so easily as a clay loam.

Some of the texture types in the texture triangle have been deliberately omitted from this classification, but they can be interpreted as intergrades between the ones which have been given.

Effects of stones and gravel

Stones and gravel, being greater than 2 mm diameter are not included in soil texture assessments. They can, however, greatly modify soil properties if present in large quantities.

Gravel is usually associated with sandy soil. In fact much of the so-called sand harvested from sand quarries should, by strict definition, be classed as gravel. There is no well-defined upper limit to gravel size.

The presence of appreciable amounts of gravel will exacerbate the problems of excess drainage and leaching found in sandy soils. Amounts of gravel in other texture types will be small.

Stones can occur in quite large proportions, particularly in the glacial soils of northern Britain. In some fluvio-glacial soils, stones can occupy 70 per cent of the volume of solid matter and, even in glacial till of clay loam texture, some 20–40 per cent of stone can occur. Stones modify soil problems in several ways:

- There is less rooting space.
- The water-holding capacity of the soil is reduced.
- Stones heat up more quickly than moist soil. Thus for a given soil texture a stony soil will be warmer than a stone-free soil.
- Effects on cultivating and harvesting implements can be severe, especially if the stones are large and angular.
- Damage to crops such as potatoes can occur at harvest.

Effects of organic matter and calcium carbonate on texture assessments

If soil contains more than 20 per cent of either organic matter or calcium carbonate there is little point in assessing the texture by laboratory methods. When present in smaller quantities, both organic matter and calcium carbonate affect field tests for texture. Moderate quantities of organic matter tend to make the soil feel 'loamy'. Well-decomposed humus feels somewhat like silt but is usually distinguished by its dark colour and its tendency to break up when smeared on the palm of the hand, which silt will not do. Moderate quantities of fine calcium carbonate also tend to make the soil feel silty.

However, in the vast majority of cultivated soils in the British Isles the quantity of organic matter is not sufficient

to affect texture assessments to any great extent. In naturally non-calcareous soils the same would be true of calcium carbonate applied as a liming material.

Effects of texture on productivity

Soils containing 15–30 per cent clay, 40–55 per cent sand and 15–30 per cent silt (some loams, sandy clay loams and sandy loams, Fig. 6.1(a)) will present few problems in cultivations, drainage or availability of water and nutrients. As the soil texture approaches any corner of the texture triangle problems of various kinds will arise. Table 9.1 (page 133) gives a general summary of problems that may be expected. There will be some exceptions to the picture presented, especially if the effects of texture are modified by high humus content but the statements will be generally valid.

Modification of texture

Soil texture is an inherent property and changes extremely slowly except in the case of some catastrophe such as massive erosion. There is little the farmer can do to change the texture of his soil in these days of expensive labour and fuel. In the days of cheap labour it was possible to apply large quantities of clay or marl, which is to sandy soils to give them some 'body'.

For marling to be feasible today would require a nearby source of marl and heavy earth-moving equipment to add some 100–200 t/ha of the material. Obviously the cost of such operations would be very high and could be justified only if large financial returns were expected in intensive cash-cropping areas. The problem of lightening heavy soils is even more difficult and would need the addition of some 700 t/ha of coarse sand, gravel or coarse boiler ash. This has been done on some horticultural holdings but is economically out of the question in agriculture.

Although the farmer must normally accept the texture of the soil as a lucky or unlucky inheritance, there is much he can do to modify the adverse effects of excessively sandy, clayey or silty soils by building up or maintaining organic matter and maintaining as good a soil structure as can be achieved under his farming conditions.

Structure

Soil structure is a fascinating and critically important soil property controlling the availability of water to plants and the ability of fine feeding roots to ramify within the soil and hence to exploit the essential plant nutrient supply.

The simplest structural units occur on newly deposited sand dunes, raised beaches and other very young soils in which there has been no time for structure development.

In these very immature soils the textural units are also the structure units. They are often said to be structureless but, in fact, have a single-grain structure.

Structure development

As the soil develops, organic matter is incorporated and humified, wetting and drying occur, plant roots colonize the soil, the surface soil freezes and thaws periodically. As a result of these and other factors soil structure forms by the aggregation of textural units. Structure can vary from rounded units of 1–2 mm diameter to large prisms 50 cm in length and 10 cm across.

The profile of a clay soil under well-managed permanent pasture can show a very clear structure pattern. This is illustrated in Fig. 6.2. The structure units in the A horizon are small (1–5 mm), rounded and porous. With depth they give way, in the B horizon to larger, more angular units and in the C horizon to prisms between which there are deep vertical fissures (Fig. 6.3).

The C horizon, a metre below the soil surface contains very little organic matter and only sparse roots penetrate between the prismatic units. The main structure-forming influence in this horizon is the swelling and shrinking of the clay fraction resulting from seasonal wetting and drying. During the summer the soil shrinks and cracks into prisms between which roots penetrate and cause further

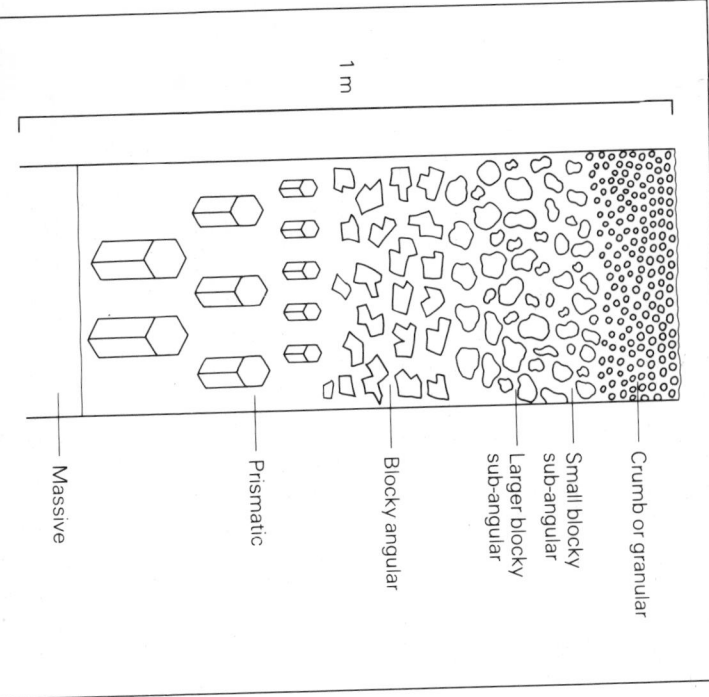

Figure 6.2 The structure of a clay soil under well-managed permanent pasture

Massive	
Prismatic	
Blocky angular	
Larger blocky sub-angular	
Small blocky sub-angular	
Crumb or granular	

1 m

Figure 6.3 A clay soil with prismatic structure

drying out. Roots cannot penetrate the very fine pores within the prisms. In autumn or winter the fissures tend to close as the soil wets and swells. Air is eliminated and roots are suffocated.

Nearer the surface, in the B and lower A horizons, the soil tends to wet and dry more frequently. This causes greater strains on the clay fabric and the soil breaks into smaller aggregates. Other influences also operate. Frost penetrates and the process of ice formation within the soil causes new pressures that are absent at depth. The combined result of these influences is a range of aggregates formed mainly by differential pressures and called 'blocky'. Immediately above the prismatic horizon blocky structure units will be large (20–50 mm), angular and still difficult for roots to penetrate (Fig. 6.4).

Nearer the surface yet, the influence of organic matter, roots and earthworms becomes important and units become 'sub-angular', with rounded off corners. These units contain larger pores which roots can penetrate more easily (Fig. 6.5). In this zone, where the A and B horizons merge, the essential change in structure-forming factors is the takeover of biological from physical influences. Roots penetrate and ramify more vigorously. These roots die and contribute organic matter which is converted to humus. This supplies to the soil, by bacterial action during humification, gum-like polysaccharides which bind aggregates together but leave pores within them which

Figure 6.4 Angular blocky structure

Figure 6.5   Sub-angular blocky structure

Figure 6.6   Crumb structure

hold available water and into which roots can penetrate.

Organic matter deposited on the soil surface as leaf litter is incorporated into the soil by earthworms. During this process, rounded aggregates described as 'crumb' and 'granular', are produced and dominate the A horizon (Fig. 6.6). They contain adequate air-filled pores and have a high available water capacity.

Once these biological influences have become established they have cumulative beneficial effects on plant growth. The good pore-size distribution encourages rooting which leads to better nutrient uptake, giving better plant growth and a greater return of organic residues, which gives increased earthworm and microbial activity and further improvement in structure.

Factors which affect soil structure

Several of the factors which affect the structure of undisturbed soils have been illustrated in the previous section.

Active humification is important in the formation of good soil structure. Well decomposed humus is porous and forms strong linkages with clay which lead to the formation of stable crumb and granular structures. Micro-organisms, during humification produce organic binding materials and

earthworms excrete pellets in which organic and mineral particles are bound together.

Plant roots exert pressures within soil. Especially important are the effects of the intimate root systems of grassland.

Wetting and drying are particularly important in clay soils which form blocky and prismatic structural units by shrinkage in dry periods.

Freezing and thawing cause pressure and release of pressure, respectively, because of the expansion of water when it freezes. This can cause the compression of soil to form structural units. It can also result in 'frost heaving', the disruption of the surface of the soil with the breaking of plant roots.

Free iron oxides present in some soils give rise to stable crumb or granular structures.

Liming affects soil structure directly through the flocculation of clay colloids to form aggregates and, indirectly, lime is essential to the survival of earthworms and to the activity of humifying bacteria. The same effects occur in naturally calcareous soils.

Fertilizers, along with lime, help to produce large crop

residues which contribute to humus formation and hence to good structure.

Manures act similarly to fertilizers but also contribute directly to the humus content of soil.

Cultivations disturb and disrupt the naturally formed structures and create temporary structure units in the production of a tilth. Excessive cultivations help to remove the binding organic matter from soil and cause structure deterioration. Above all, cultivation of heavy soils when wet can bring about platy structure, horizontal laminar layers, in the form of cultivation pans.

Drainage operations, if done with skill, reduce waterlogging, increase biological activity and improve soil structure.

Classification of structure

Table 6.2 gives a classification of soil structure units which is easily applied to British conditions. The descriptive terms convey the general properties of the structure units very well. The table also gives some examples of conditions under which the structure types will be found and some of their properties. The organic matter contents given are intended only as a general guide. If there is some

overwhelming influence on structure, such as a high iron oxide content in the soil, the organic matter figures will not apply.

'Good' and 'poor' soil structure

'Good' soil structure is firmly bonded, stable, allows intimate root penetration, drains away excess water readily but retains a maximum of available water for the plant. Such structure is encouraged by lack of disturbance, a continuous supply of organic matter, good drainage and neutral or mildly acid pH – conditions fulfilled by well-managed pasture on medium-textured soils. The structures formed are crumb, granular and to a lesser extent blocky sub-angular.

'Poor' structure types include massive, prismatic, and angular blocky. They are encouraged by low organic matter content, heavy texture, poor drainage and lack of lime. Platy structure is also poor because it restricts water movement. It is caused mainly by inept cultivations (Fig. 6.7).

Permanence and stability of structure

No type of soil structure can be regarded as truly permanent. Even the best crumb structures are dependent on a long-established equilibrium in the soil. If this is

Table 6.2 Classification and description of soil structure units

Unit size and description	Examples of occurrence	Main properties
Single grain	Young soils, e.g. dune sands, raised beach, wind-borne silts. Leached grey horizons of Podsols	Excessive drainage. Surface soils subject to wind erosion Very low organic matter (less than 1%)
Granular Rounded aggregates Small: 1–2 mm Medium: 2–5 mm Large: 5–10 mm	Surface soils, especially medium textured, formed beneath old broadleaved forest or well-drained poor permanent pasture. Some cultivated surface soils	Porous. Structure units penetrable by fine roots. Good air/water balance. Excess water drains readily. High available water capacity Moderate organic matter (3–5%)
Crumb Rounded aggregates 2–5 mm	Similar to granular but occurring beneath vigorous broadleaved forest with rich ground flora or excellent, well-limed permanent pasture. Some cultivated surface soils	Similar to granular but with beneficial properties enhanced High organic matter (5–10%)
Blocky sub-angular Angular aggregates with rounded corners Small: 5–10 mm Medium: 10–20 mm Large: 20–50 mm	Subsoils lying beneath granular or crumb-structured topsoils. Some cultivated surface soils, e.g. after three-year leys on medium and heavy soils	Structure units more difficult to penetrate by fine roots than crumb or granular. Availability of water to plants dependent on texture – good in light-textured soils, moderate in heavy soils. Hard when dry Moderate or low organic matter (2–4%)
Blocky angular Angular aggregates Small: 5–10 mm Medium: 10–20 mm Large: 20–50 mm	Subsoils lying beneath blocky sub-angular topsoils. Topsoils of some heavy-textured soils. Commonly associated with imperfect or poor drainage	Fine roots penetrate structure units very little. Availability of water to plants adversely affected by lack of root penetration if units are large Organic matter low or very low (less than 1–3%)

Unit size and description	Examples of occurrence	Main properties
Prismatic Prisms with vertical axis 5–6 times as long as the horizontal cross-section. Size variable. Cross-section in some subsoils can be as much as 80–120 mm	Heavy-textured subsoils. Commonly associated with poor drainage and marked seasonal wetting and drying	Fine roots cannot penetrate structure units. Roots grow between units in dry periods. Availability of water to plants poor, even when soil is almost saturated with water Organic matter very low (less than 1%)
Platy Can be continuous horizontally orientated compacted horizon. Horizontal axis much greater than vertical thickness. Thin: less than 2 mm Medium: 2–5 mm Thick: 5–10 mm	Occurs naturally in some Podsol horizons. More commonly caused by cultivation of medium-heavy textured soils when plastic. In this case pans with platy structure occur immediately beneath the plough layer	Very compact. Roots cannot penetrate and can form a horizontal mat along the upper surface. Water percolates either not at all or with difficulty Organic matter variable, but usually low
Massive A continuous mass of soil with no cracks	Heavy-textured subsoils in areas of high rainfall and cold climate	Usually waterlogged. Roots do not penetrate Organic matter very low (less than 1%)

disturbed by cultivation, serious losses of organic matter will occur. As a result the crumb structure will deteriorate rapidly within a few years or even months.

Some types of structure have no persistence – for example those formed by drying and shrinkage which will 'slake' and disintegrate under heavy rain.

It is important to maintain, where possible, a type of structure which will be stable against the onslaught of water falling as rain on the surface or passing through the soil. This can be achieved in many soils by building up organic matter, liming, draining and disturbing the soil as little as possible. Unstable structures result in surface 'puddling' and 'capping' in which thin impervious layers are created by rain splash from slaked structure units at the soil surface.

Another result of structure breakdown, but in this case within the soil, is the downward movement of deflocculated, slaked, clay particles through the 'sieve' of larger particles or aggregates. The clay may be deposited lower in the profile and this produces a horizon which is described as a 'textural B horizon' or as a 'clay pan'. These horizons are widespread in Podsols and Brown Earths in the British Isles. As it deposits, the clay forms 'skins', which can be seen with a lens, on the particles of the B horizon and the drainage routes become blocked. These clay pans are dense and can prevent the passage of both

Figure 6.7   Platy structure

water and roots. In cultivated soils they usually lie below the cultivated layer.

Tilth

Tilth is created by cultivating the soil and thus breaking up the clods left by ploughing to a size and condition which will encourage the germination and growth of crops.

The production of tilth is an art practised with great skill by some farmers. The judgement required is such that one or two days' anticipation or delay can have serious effects on seedling emergence. The greatest temptation, especially in wet, cold springs is to cultivate before the soil is dry enough, while it is still plastic. This results in compaction.

What the farmer is doing in creating a tilth is to produce artificial, temporary and often unstable structure units. On many types of soil his skill will carry the crop through the critical emergence stage, after which the roots of the crop will help to create or maintain structure. Many light- and medium-textured soils form a tilth very easily. It is on sandy clay loams, clay loams, silty clays and clays that difficulties arise. The aim should be to produce a range of aggregates between 2 and 20 mm in size at the surface without compacting the soil below.

It is unfortunate that the intensive cultivations necessary to produce a tilth are detrimental to the maintenance of satisfactory structure. Equally unfortunate is the fact that the poorer the natural structure the more intensive are the cultivations required to create a good tilth. These are some of the reasons why many farmers are looking into the possibilities of direct drilling or minimal cultivations. The role of these systems in structure maintenance is described in Chapter 10.

Soil water and air

7

Both water and air are essential for plant growth. The plant also absorbs most of its nutrients from the soil in water. Water and air occupy the pore spaces between the organic and mineral matter. If water moves in air moves out and vice–versa. The total pore space of a well-structured topsoil can occupy 50–60 per cent of the volume whereas only 25–30 per cent of a compact subsoil or plough pan will consist of pores.

The larger pores (macropores) of a well-drained soil are normally occupied by air which can move freely in and out of the soil and so maintain the oxygen necessary for plant roots to thrive. Immediately after rain these pores may be temporarily full of water until it drains away. If, however, drainage is impeded or if the soil is heavy textured and located in a wet area, all the pores will fill with water and air will be excluded. The soil is then said to be waterlogged. Anaerobic conditions will develop with consequent ill-effects such as the production of toxic gases, the asphyxiation of roots and the death of plants.

The amount of water that the soil can retain and hold readily available to plants is determined by the size distribution of individual pores and especially the proportion of small pores (micropores). All soils can retain some water in these pores and there is a constant tug-of-war between the forces retaining water in the soil and those trying to remove it upwards (transpiration of plants and evaporation from the soil surface) and downwards (gravity).

The water retained by very small pores and in very thin layers around soil particles is very strongly held and plants cannot exert sufficient suction to withdraw it. This water is unavailable to the plant. The critical suction that plants can exert is constant for all species and is approximately 15 bar (equivalent to 15 times atmospheric pressure). If water is held by forces greater than this the soil is, so far as the plant is concerned, dry.

Water is held in some pores at less than 15 bar tension and is, therefore, capable of being absorbed by active plant roots. This water can move by capillary action, as through blotting paper, to meet the root system. If the soil is moist enough, aerobic enough, and is supplying adequate nutrition, the plant roots will actively move towards available water. All the processes of water gain and loss, water movement and root movement are strongly affected by soil texture, structure and humus content. These processes determine the plant–soil water relationships on which crop yields are probably more dependent than on any other single factor.

Water requirement of plants

Plants need to take in through their roots and transpire 200–500 units of water for every unit of dry matter produced. This means that a grain crop yielding 10 t/ha of dry matter (grain and straw) will transpire 2000–5000 t/ha of water. This alone would account for rainfall equivalent to 200–500 mm (about 8–20 in) and this makes no allowance for drainage losses, surface run-off and evaporation from the soil surface. It is easy to see that in the drier southern and eastern parts of the British Isles, lack of water will frequently reduce crop yields. Even in areas of relatively high rainfall crop growth can be severely restricted on many types of soil by lack of water over relatively short periods and conservation of water is of major importance.

Water retention and availability

Consider a soil which has been saturated with water by continuous rain or irrigation. It will have a glistening appearance and all the pores will be full of water. Air is temporarily excluded. Provided that the drainage is not impeded some water (in pores wider than about 3 mm) will drain away rapidly and will be followed more slowly by water from somewhat smaller pores (approximately 30 μm–3 mm). This so-called 'excess' water is of little use to plants as it is transient in well-drained soils and restricts aeration in poorly drained soils. The pores it occupies are known as macropores.

After all excess water has drained away the soil is said to be at field capacity. This is the normal state of a well-drained soil before plant growth has begun in the spring. At this stage, all the small micropores (less than 30 μm) are full of water and the macropores are full of air.

'Available water' is held in the micropores by capillary forces and can move in any direction in the soil.

As soil containing available water dries out as a result of plant growth or drought a stage will be reached at which there is no more water available for plants. Sensitive crops such as sugar beet will wilt in the daytime and recover at night, probably because water is slowly reaching the roots by capillary movement. If drying continues, a point is reached at which plants will not recover. This is known as the 'permanent wilting point' (15 bar tension), beyond which the plant cannot exert sufficient suction to remove any further water from the soil. All but the very finest pores are then full of air, but there is still 'unavailable water' held at tensions greater than 15 bar.

Because this water is held mainly in very thin films around soil particles, the amount of it is strongly influenced by the surface area of the particles. In a clay soil this is an extraordinary 20–25 ha/kg of soil compared with 3–5 ha in a sandy soil. Thus clay soils contain much more unavailable moisture than sandy soils, sometimes as much as 15 per cent of their own dry weight (equivalent to almost 40 mm of rainwater distributed through the plough layer).

Effects of soil texture   Figure 7.1 shows the tension which must be overcome to remove water from two soils (sandy loam and clay loam) with various water contents. The horizontal scale is logarithmic so that each division represents a 10-fold increase in tension. The important zone is that between wilting point and field capacity, representing 'available water'. The plant root needs to exert more than 15 times as much suction to remove water at wilting point as compared with field capacity. Plants are able to do this and there is no apparent restriction of water uptake until the wilting point is very near (10–15 bar tension). In these conditions it is probably the speed at which water can move through the macropores into zones dried out by plant roots and not the ability of the roots to exert sufficient suction that controls water uptake.

Figure 7.2 shows the general relationship between soil texture, available water capacity and unavailable water capacity, assuming a standard organic matter content. Both organic matter and soil structure type can modify this relationship – good structure and high organic matter tend to increase the available water capacity.

The capacity of the soil to hold unavailable water increases steadily with clay content (Fig. 7.2), but the available water capacity reaches a maximum in the middle range of texture and can be as high for a loam as for a clay soil.

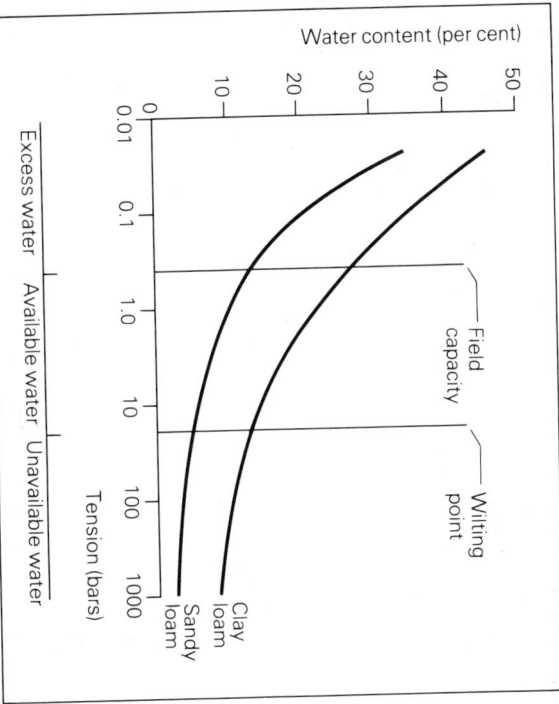

Figure 7.1 The relationship between soil texture, water content and water tension

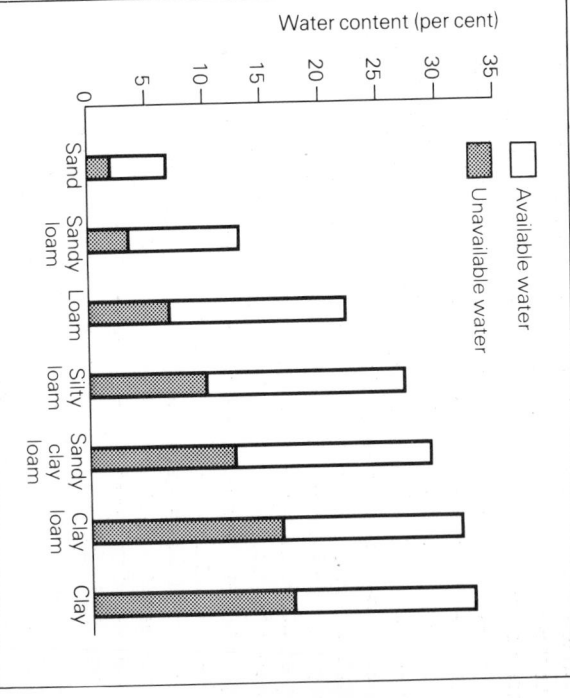

Figure 7.2 Soil texture, available water capacity and unavailable water capacity

In Fig. 7.2, the available water capacity varies from only 5 per cent for the sand to 17 per cent for the silty loam. Assuming the weight of topsoil (25 cm deep) to be 2500 t/ha, the storage capacity for available water of the sand is only 125 t/ha and that of the silt loam is 425 t/ha. These are equivalent respectively to 12.5 and 42.5 mm of rain. Any water in excess of this must occupy the macropores or pass downwards into the subsoil.

Effects of soil depth   The more subsoil water that plants can draw upon the better. The depth to which roots can penetrate depends upon the plant species, soil structure and the availability of water at depth.

The roots of some crops, such as cereals and lucerne can penetrate well-drained soils to great depths, as much as 150 cm, but this is uncommon and certainly the intensity of the root system declines rapidly with depth. Other crops such as sugar beet and potatoes draw most of their water from the surface soil. None the less, many crops can draw appreciable amounts of water from as far as a metre below the soil surface and any pans or waterlogged horizons that restrict root penetration will interfere with this.

Effects of organic matter and structure   The water retention of all texture types can be considerably affected by well-decomposed humus. Figure 7.3 illustrates the relationship for a soil of loam texture. The increase in unavailable water capacity is mainly a result of the absorptive properties of colloidal organic matter which holds water tightly. There is a much more important increase in available water capacity from 10 per cent for the soil containing 1 per cent of organic matter to 14 per cent for soil with 6 per cent organic matter. This is a result not simply of the absorptive properties of humus but also of the beneficial effects on soil structure with consequential increases in the volume of micropores within the structure

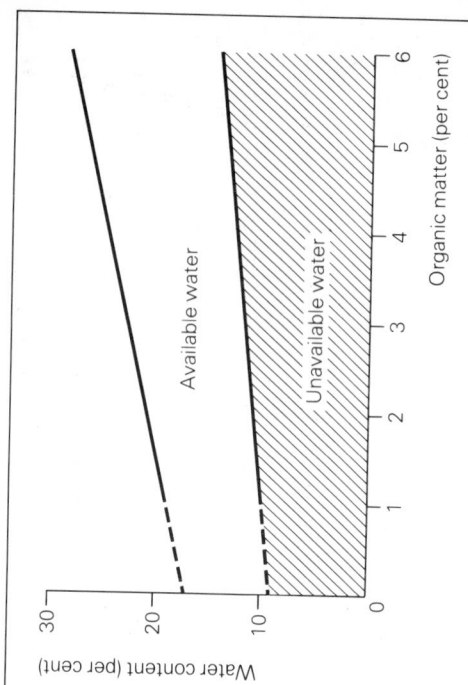

Figure 7.3   An example of the effect of organic matter on the water-holding capacity of a loam

units. Such an increase in available water capacity can prevent the percolation of some 10 mm of rain from the topsoil where most of the plant roots are concentrated.

The effect of organic matter on available water capacity will be *proportionally* much greater in light-textured than in heavy-textured soils. In fact, in sandy soils, the only effective way of increasing water-holding capacity is by increasing the organic matter content.

Movement of water in soil

In order to function efficiently a plant root needs to be close to a zone which contains available water. As it takes up water the plant root dries the soil immediately next to it to some extent. This is illustrated by the formation of tubes of ferric oxide around roots in some types of Gley soil. If the soil water tension immediately around the root is brought to wilting point, the root will soon cease to function. This can be prevented only by the extension of the root into new zones where water is held at lower tensions or by movement of water into the depleted zone before the root dies.

During periods of vigorous plant growth the root system extends and ramifies rapidly and taps available water supplies effectively. On the other hand, in periods of slow growth, whatever the cause, the movement of water to the root becomes more important.

Water moves in soil by two influences: by gravity downwards and by capillarity in any direction. Capillary movement of available water is always from wet zones (low tension) to dry zones (high tension). It moves rapidly in the larger range of micropores (10–30 μm) and very slowly in the smallest micropores (less than 1 μm). Therefore, movement is most rapid in sands, loamy sands and sandy loams but can be very slow in clays and clay loams.

Water can move upwards by capillarity from a water-table. This happens in Ground-Water Gley soils. Immediately above the water-table is a zone enriched with available water. In sands and sandy loams this zone is 30–40 cm and in clays and clay loams 60–80 cm in depth. This zone of available water can have important benefits to crop production in a wide range of Gley soils with water-tables 1–1.5 m from the soil surface. If depleted of water by root action this zone will be replenished, in a sandy soil within a day. In a clay soil replenishment may take a week or two.

During the growing season equilibrium is seldom reached in the micropores and available water is constantly on the move in response to root action.

Trapped available water

In some cases water which is theoretically available to the plant in that it is held at tensions lower than 15 bar cannot be taken up because it is 'trapped' in a part of the soil where it cannot be absorbed by plant roots.

Figure 7.4 illustrates the action of roots in three cases where trapped water occurs. Prismatic structure (Fig. 7.4(b)) is usually found in the lower horizons of heavy-textured soils. In spring, when the soil is wet to field capacity, these horizons may be waterlogged and roots cannot penetrate. As the soil dries out roots can penetrate

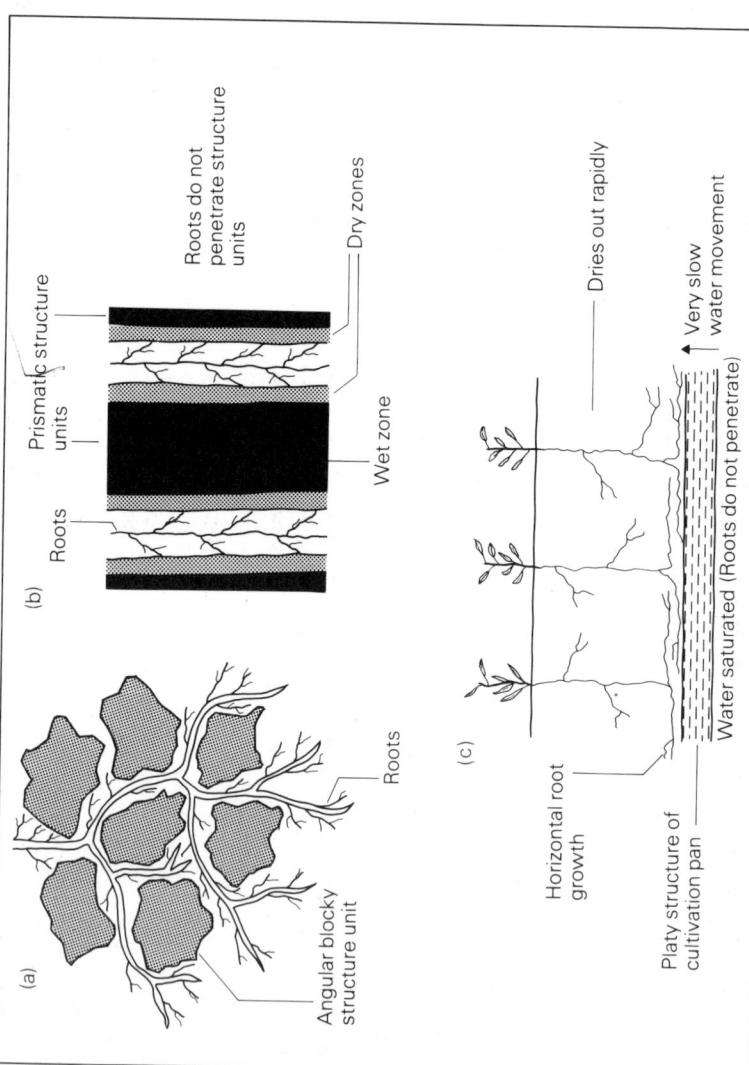

Figure 7.4 Root penetration and water availability

only between the large structure units and, because of the preponderance of fine micropores, cannot enter the prism itself. In a period of active plant growth the zone of easy access of water to roots on the edges of structure units is rapidly depleted. Water cannot move quickly enough from inside the prismatic unit and a dry zone is established around its edges. Such soil can have a total water content indicating ample available water but it is positionally unavailable.

The case of the poorly cultivated clay soil with large clods on the surface (Fig. 7.4(a)) is similar but the condition is exacerbated by the action of sun and wind which dries the surfaces of the clods. If these apparently dry clods are broken a discrete dry layer can be seen around the edges but the soil in the inner parts is wet enough to be moulded.

The young roots of newly germinated seedlings have a very poor environment in which to develop. They are surrounded by very large pores and structure units, dry on their surfaces, which are difficult to penetrate.

Figure 7.4(c) shows the effects of a cultivation pan. The problem in this case is brought about by compaction which reduces the total pore space and increases the proportion of fine micropores through which water moves very slowly. Such pans can severely restrict the amount of water received by the plant, both by preventing the passage of roots through them and by retarding the upward movement of water to meet the roots. In fact many such pans, lying immediately below the plough layer, can deprive the plant of access to all lower horizons and the surface soil, in these circumstances, dries out very rapidly. Other types of pan (iron pans, clay pans, hard pans) have similar effects to cultivation pans.

Effects of large voids

In sharp contrast to the restriction of water movement by pans is the very rapid passage of water through the networks of large voids in some soils. In Brown Earth soils with a high earthworm population, worm holes 3 or 4 mm in diameter may occupy a large part of the soil encouraging very rapid loss of excess water. In some Gley soils the large structure cracks between prismatic units after a dry spell can be a centimetre or more in width and water will pour down them in the event of heavy rain – often before it has time to be absorbed by the topsoil. Thus the 'model' homogeneous soils used by some soil scientists to predict soil–water relationships are of very little value.

Removal of water from the soil

Water is removed from the soil downwards by percolation and drainage and upwards by evaporation from the soil surface and by transpiration. There is also a proportion of precipitation (rain, snow and small contributions from fog, hail, dew) which does not enter the soil but is intercepted by vegetation and evaporated into the atmosphere. Other water reaches the soil surface but does not infiltrate, being lost in surface run-off.

Percolation and drainage

More than half of the precipitation falling on bare soil in the British Isles can be lost in drainage. Drainage losses in winter can be twice as great as those in summer because of evaporation from the soil surface in the warmer summer. This picture changes radically if the soil carries a crop. If the soil is assumed to be at field capacity (containing no excess water but maximum available water) in March, an intensively managed grass crop yielding 12–15 t of dry matter per hectare over the season can be expected to reduce drainage losses to negligible amounts during the period of April–September unless the distribution of rainfall is very bad, with heavy periods of rain followed by dry spells. Annual crops, particularly row crops do not cut down drainage losses nearly so much. Grass starts the season with an active root system but a crop such as potatoes may be planted in March or April in wide drills and will transpire very little water in the first 6–8 weeks of growth when the plants are growing slowly. In this case

appreciable drainage losses of precious water may occur in April–June after which there is a period of vigorous growth and drainage losses will be small.

Surface run-off

Some rainfall unfortunately never enters the soil but runs off the surface downslope or forms puddles on the surface. Surface run-off can be very serious in some parts of the world, accounting for more than half the annual precipitation and causing erosion of valuable topsoil. It is particularly bad when rain is torrential, fortunately rare in this country. Surface run-off occurs because water is unable to infiltrate, that is to pass through the surface of the soil quickly enough. It is most serious on steep slopes on heavy soils in high rainfall areas, which are so wet and impermeable that water cannot enter the surface soil. Run-off will also occur on soils with unstable structure in the surface horizon often associated with very low organic matter content.

Evapo-transpiration

Evapo-transpiration covers all water vapour losses from soil and plant surfaces – evaporation from soil, transpiration by the plant and direct evaporation of water intercepted by the plant leaf canopy. The last mentioned cannot strictly be regarded as 'removal of water from the soil' but is considered here as an essential part of the cycle of water between soil, plant and atmosphere.

Evaporation from the soil surface is great in semi-arid regions and can account for three-quarters of the annual rainfall in such areas. Evaporation is greatest from bare soils, is accelerated by warm winds, but is reduced to a minimum by complete plant cover. It is much smaller in cool humid conditions but can still account for 20–60 mm of water per year on soil which is left bare for long periods.

Interception and evaporation of water from the leaf canopy must be accepted as an inevitable loss. In the deeply layered canopies of broad-leaved forest trees, interception of rainwater is so great that showers of 2–3 mm may never reach the soil. Interception by agricultural crops is much less, being negligible in the early stages of growth of row crops, highest on grass and quite small in smooth-leaved crops such as sugar beet.

Transpiration losses are both unavoidable and desirable so long as the transpiring plant is part of the commercial crop. Unfortunately weeds also transpire and by far the most serious aspect of weed competition is that of reducing the amount of water available for the crop.

The number of units of water that a crop transpires to produce one unit of plant dry matter is called the transpiration ratio. There is a very simple approximate relationship between transpiration ratio and the amount of water (rainfall equivalent in millimetres) required to produce 1 t of plant dry matter *per hectare*. The water requirement is one-tenth of the transpiration ratio (transpiration ratio 200, water requirement 20 mm). Thus to produce a crop of barley (grain plus straw) of 10 t/ha would require, at a transpiration ratio of 200, some 200 mm of water. Similarly a crop of 50 ha of potatoes at the time of maximum dry matter production (tubers plus haulms) would require, at a transpiration ratio of 500, some 2500 mm of rain equivalent – double the annual rainfall in the drier parts of the country. The fact that such a crop can be grown without irrigation is a simple indication that transpiration ratios are towards the low end of the range in the climate of this country. Experimental work indicates that the ratio in the British Isles is 200–250. The higher range of ratios (400–500) is found in warmer climates with lower humidity.

None the less the simple examples taken illustrate that water supply must frequently be the limiting factor to crop growth, especially if it is remembered that these calculations take no account of drainage, run-off and losses by evaporation from the soil surface.

The greatest efficiency of use of transpired water – that is, the lowest transpiration ratio in a given locality, is found where excellent crops are being produced by a combination of good farming practices – drainage, cultivations, lime and fertilizer use. If any one of these is neglected more water will be needed to produce 1 unit of dry matter.

Aeration

A well-aerated soil is one in which the essential oxygen is available to plants and other aerobic organisms in sufficient quantities for optimum growth. At the same time carbon dioxide, being produced in the soil by root respiration and by aerobic decomposition of organic matter must not be allowed to build up, as this will adversely affect plant growth. Poor aeration occurs when soil water occupies all or most of the pore space. It can also occur when there is apparently sufficient total air space. The movement of gases to and from the soil is directly related to the proportion of air-filled pores. If there is a continuous chain of air-filled macropores, as in well-drained sandy soils, there are few problems. If there is a large proportion of fine micropores the exchange of gases between atmosphere and soil atmosphere is very slow and carbon dioxide concentrations will increase. In poorly structured clay soils the exchange can be made critically slow by the slow penetration of newly fallen rain into the surface soil preventing the escape of carbon dioxide.

The total pore space of soil ranges from about 30 per cent of the whole soil volume in some sandy surface soils to 60 per cent or more in well-structured loams, clay loams and clays. It is seldom the total pore space that restricts aeration but rather the proportion filled by water and the rate at which air can diffuse in and out.

As a general rule a soil in which the ratio, by volume, of air to water is 1 or more will have no aeration problems. Figure 7.5 illustrates the proportions of the soil volume occupied by solid particles, water and air in two soils adversely affected by agricultural operations. Figure 7.5(a) represents a sandy loam soil, ploughed regularly to a depth of 20 cm, which is suffering from both surface compaction and a plough pan. Despite the reduced volume of air-filled pores at 0–5 and 20–25 cm below the surface there would be sufficient oxygen for plant growth.

Figure 7.5(b) represents a sandy clay loam under pasture, previously compacted at depth and recently seriously 'poached' (compacted at the surface) by the hooves of cattle. In this soil the air-filled pore space in the surface 8 cm is only 1.2 per cent and anaerobic conditions are inevitable.

Adverse effects of poor aeration on crops

Many of the effects associated with temporary or permanent waterlogging are caused by lack of oxygen. This can result in severe crop damage, sometimes after only one

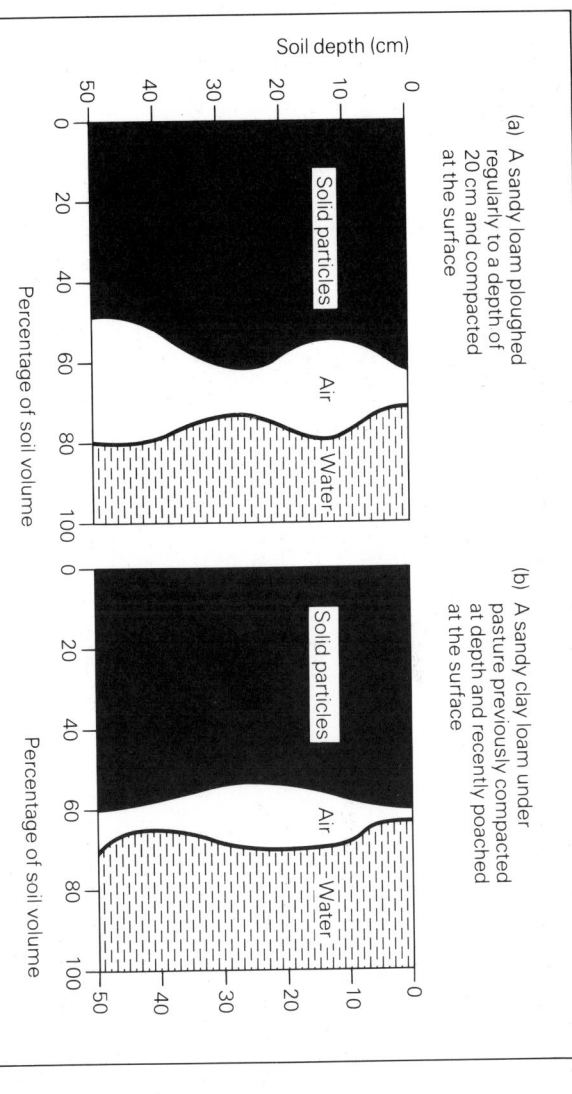

(a) A sandy loam ploughed regularly to a depth of 20 cm and compacted at the surface

(b) A sandy clay loam under pasture previously compacted at depth and recently poached at the surface

Figure 7.5 Effects of cultivations and stocking on the volume occupied by air, water and solids

or two days of flooding. Unfortunately, plants that have started their life in well-aerated soil are more adversely affected by subsequent poor aeration than those which have been poorly aerated in the early stages of growth.

There are some crops such as paddy rice which can grow well with very low oxygen supplies but it is surprising that some grass species can tolerate long periods (20–30 days) of flooding. Most clovers seem to tolerate 10–20 days of flooding but some root crops such as sugar beet can be adversely affected within a few days and most susceptible of all are cereal crops. Even short periods of heavy rainfall can adversely affect cereal crops grown on poorly structured clay soils.

Changes resulting from poor aeration

Soil oxygen may become completely exhausted. This completely changes the nature of the active microbial population. The aerobic bacteria, fungi and other organisms involved in organic matter decomposition in well-aerated soil are suppressed and anaerobic organisms take over. The results are very important, particularly if new organic matter such as farmyard manure or straw is added to the soil.

Toxic gases such as hydrogen sulphide (H_2S), methane (CH_4) and ethylene (C_2H_4) are formed by the action of anaerobic micro-organisms. Very small concentrations of ethylene can restrict plant growth.

Carbon dioxide may build up to levels at which it actively restricts the use of oxygen by plant roots. There is 0.03 per cent of carbon dioxide in atmospheric air and always very much more (8–300 times as much) in soil air. When carbon dioxide levels exceed 1 per cent adverse effects can occur.

Denitrification occurs, a process during which nitrate nitrogen (NO_3^-), from fertilizers or produced from organic matter during aerobic periods, is converted to elemental nitrogen or reduced nitrogen compounds such as nitrous oxide (N_2O). Nitrogen in these forms is unavailable to the plant. There is strong evidence that losses of available nitrogen in this way can be very serious if anaerobic conditions persist in the soil for more than a day or two.

Earthworms and other soil animals, which eat organic matter and mix it with mineral matter, are asphyxiated with consequent ill-effects on soil structure.

Reduction of iron and manganese compounds occurs giving the ferrous and manganous forms which, particularly in acid conditions, become toxic to plants. Another result of the reduction of iron compounds is the grey or blue-grey colours which dominate in anaerobic soil horizons.

Thus the presence of adequate amounts of air, low in carbon dioxide content, in soil at all times and the maintenance of its oxygen content are critical to good humification, good soil structure, availability of nitrogen, the prevention of accumulation of toxic substances and consequent good plant growth.

Aims of soil water management

Good soil water management will also ensure adequate aeration, as it will include the rapid removal of excess water from macropores.

Unless irrigation is to be used the farmer has virtually

no control over the amount of water received by the soil. The texture of the soil, which also has a large influence on the amount of water available for transpiration by crops, is also outwith his control, but several other aspects of soil water balance can be influenced by good husbandry:

- Losses by evaporation from the soil surface should be minimized by early crop establishment and maintenance of soil cover by crops or mulches.
- Weed competition should be minimized.
- Retention of excess water in the soil should be reduced or eliminated by suitable drainage and subsoiling operations.
- The formation of cultivation pans, by working heavy soils when too wet, should be avoided.
- The available water capacity should be maintained or improved by avoiding structure breakdown and by returning all possible organic matter to the soil. This will also improve root penetration and ramification and reduce the leaching of nutrients.
- Cultivations, in heavy soils, should aim to reduce clod size in order to improve root penetration.

Plant nutrients

8

This chapter deals mainly with factors affecting the availability of nutrients to plants. Practical aspects of the occurrence, prevention and correction of nutrient deficiencies in the field are considered in Chapter 16.

Elements and their functions

The soil is the prime source of essential elements (nutrients) for plants. The only exception to this is carbon, which plants assimilate from the atmosphere in the form of carbon dioxide. Although plants are capable of absorbing some nutrients through the leaves, virtually all uptake is through the roots.

Table 8.1 shows the elements known to be essential to plant growth. There is also some evidence that other elements might be essential to plants but it is not yet conclusive. The distinction between major and trace elements simply indicates that the major elements are required by the plant in fairly large quantities (conveniently expressed as per cent or as grams per kilogram of plant dry matter) and the trace elements in relatively small amounts (normally expressed as parts per million or milligrams per kilogram of dry matter).

The absorption by the plant of insufficient quantities of any one of the essential elements can give rise to severe reductions in crop yields, in some cases indicated by

Table 8.1   Elements essential for plant growth

Major elements		Trace elements
Nitrogen (N)		Boron (B)
Phosphorus (P)		Manganese (Mn)
Potassium (K)		Copper (Cu)
Calcium (Ca)		Iron (Fe)
Magnesium (Mg)		Molybdenum (Mo)
Sulphur (S)		Zinc (Zn)
Carbon (C)		Cobalt (Co)
Hydrogen (H)		Chlorine (Cl)
Oxygen (O)		Silicon (Si)

specific symptoms on the leaves or in other parts of the plant.

Besides the elements mentioned in Table 8.1, other elements are known to be essential to animals but not to crops. Of these the more important are selenium (Se), Iodine (I), Fluorine (F) and chromium (Cr). Thus at least 20 elements are known to be essential for the growth of either plants or animals, or both.

The major elements, with the exception of potassium, are all components of substances which are part of the fabric of the plant. Nitrogen is a constituent of all proteins and sulphur of some proteins and oils. Calcium is essential at the growing points of plants and is a component of calcium pectate in cell walls. Magnesium is a constituent of chlorophyll which is essential to photosynthesis.

Phosphorus is the key to a vast number of enzyme actions within the plant. It is a constituent of the cell nucleus and is essential for cell division. Potassium in contrast to all the other major elements is not a constituent of plant fabric. Its main function in the plant is in the regulation of osmotic pressure and the maintenance of turgidity of the tissues.

The trace elements have specific functions and all of them are known to be involved in enzyme actions, acting mainly as catalysts. For this reason they are required by the plant in only very small quantities. None the less lack of a trace element can adversely affect plant growth similarly to lack of a major element.

Availability

Varying proportions of the total quantity of any nutrient element are, at a given time, in forms which can be taken up by the plant. For example some peats or soils with mor humus contain large amounts of nitrogen in complex compounds but plants cannot draw upon them. Unweathered rock minerals may contain large amounts of potassium, calcium and other elements in silicates from which the plant cannot absorb them. These are examples of chemically unavailable forms of nutrients.

Effectively absorption by the plant occurs in ionic form

from either the cation exchange complex of clay and humus or directly from the soil solution. Some nutrients are absorbed as cations (Ca^{2+}, K^+, Mg^{2+}, Mn^{2+}, Cu^{2+}), some as anions ($H_2PO_4^-$, BO_3^{3-}, MoO_4^{2-}) and nitrogen as both cations and anions (NH_4^+ and NO_3^-). Most fertilizers are designed to supply nutrients, usually restricted to N, P and K, in these readily available ionic forms.

The various pathways that can be followed by nutrient elements in soil are illustrated in Fig. 8.3, 8.4 and 8.5 for nitrogen, phosphorus and potassium. Although the patterns for the three elements vary in detail they have much in common. Figure 8.1 shows the general pathways. The nutrient ions come into the soil solution by the weathering of soil minerals, by the decomposition of organic matter such as crop residues or dung. They may then be taken up by the plant directly from the soil solution or absorbed by the cation exchange complex and subsequently taken up by the plant. Alternatively, the nutrient ions may be leached from the soil or 'fixed' in several ways in forms unavailable to the plant for either long or short periods of time. Fixation is a general term which includes temporary immobilization in the bodies of bacteria or fungi, the conversion of the ion to chemical forms of low solubility and the temporary trapping of the ion between the plate-like layers of clay minerals. Finally, the plant may be physically prevented from taking up the nutrient by barriers such as pans or waterlogged horizons.

In this case the nutrient ion remains chemically available but is positionally unavailable to the plant.

Unfortunately among the conflicting destinations for the nutrient ion the plant is commonly a poor competitor, particularly for ions such as phosphate which are subject to chemical fixation.

The factors affecting the availability of nutrients to the plant may be classed as environmental, chemical and physical.

Environmental factors

Parent material strongly influences the potential quantity of a nutrient which may become available to plants. For example sandstones will usually contain very little potassium or calcium, limestone will be rich in calcium and sometimes magnesium but will be short of potassium and many trace elements. Some parent materials are very rich in one particular element as, for example, black marine shales which contain much molybdenum. Soils derived from these fairly 'pure' parent materials will have forecastable deficiency or toxicity problems.

Other materials, such as some glacial tills contain a wide spectrum of essential nutrients in adequate quantities.

The parent material also strongly affects the texture of the soil which in turn affects the degree of leaching, so that a sandstone, already low in potassium, will form a sandy

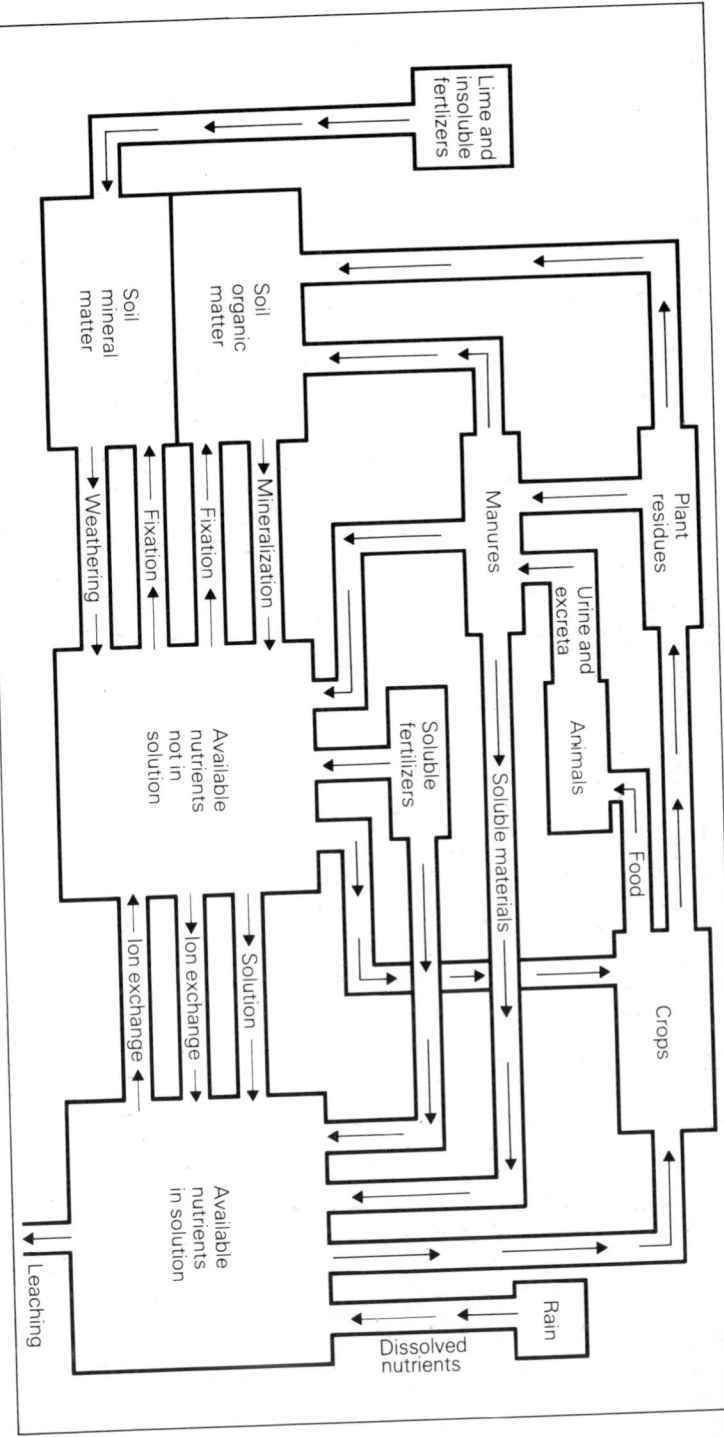

Figure 8.1 General pathways of nutrient elements in soil

soil subject to leaching of the small amounts of potassium that are weathered and enter the soil solution.

Climate is an important factor through its effects, already discussed, on the rate and type of weathering and leaching. A warm humid climate leads to rapid weathering of rock minerals, but also to high leaching often resulting in low nutrient availability. In cool humid climates weathering takes place more slowly and the availability of nutrients released by weathering depends largely on the degree of leaching brought about by local climatic conditions. Climate also strongly influences the amount of available water in the soil and this is critical to nutrient availability. The type and quantity of clay minerals in the soil also result from a combination of parent material and weathering processes.

Type of agriculture influences nutrient availability according to its intensity. In uncultivated hill soils receiving no lime or fertilizers an equilibrium is established between soil and vegetation and there is a balance between annual gains and losses of available nutrients. The natural vegetation best suited to the soil of the area will dominate and the amounts of nutrients available will be a major factor in determining this. A good example is the struggle for dominance on a hill soil between heather and bent/fescue grass communities, which is strongly influenced by the amounts of available

calcium and manganese. Heather can survive with less calcium and more manganese and will therefore tend to dominate the more acid parts of the hill.

These equilibria are very finely tuned and a period of several years of abnormal climate can cause a dominant species to become less dominant or even to be superseded. Any major interference such as reclamation for agriculture will greatly affect the equilibrium. Drainage and liming tend to deplete the soil of both major and trace elements through leaching losses. Also, because of the much higher yields brought about by fertilizers, the demand for available nutrients is much greater in commercial crops than in the natural unfertilized vegetation. If the crops are carted off there is an absolute loss of nutrients from the soil and hence the need for liming, fertilizer application and in some cases trace element sprays.

Chemical factors

A wide range of chemical factors affect nutrient availability.

Cation exchange is a process involved in several aspects of nutrient availability. It involves the exchange of cations between the clay–humus cation exchange complex, the soil solution and the plant root. The exchanges are very rapid and follow the normal laws of chemical reaction. A divalent cation such as calcium (Ca^{2+}) will exchange with two monovalent cations such as hydrogen (H^+) thus:

$$Ca\ (clay) + 2H^+ \rightleftharpoons \begin{array}{l} H \\ H \end{array}\ (clay) + Ca^{2+}$$

This equation represents the replacement of calcium adsorbed on the cation exchange complex by hydrogen ions from the soil solution. The Ca^{2+} released into the soil solution can then be leached or taken up by the plant or readsorbed on the cation complex.

Nutrients held on the complex are available for uptake by the plant either directly or by transfer to the soil solution. The amounts of various ions held in this way are strongly influenced by soil acidity, liming and fertilizer application. In acid soils there is a preponderance of hydrogen (H^+) and aluminium ions. (Al^{3+}, and other forms) In calcareous soils or after heavy liming calcium (Ca^{2+}) ions predominate. For short periods after the application of a potassium fertilizer, potassium ions (K^+) may be dominant. The dominance of the cation exchange complex by one or two specific cations is detrimental to the availability of others to the plant. For example, in a very acid soil, the complex is dominated to such an extent by hydrogen and aluminium ions that nutrient ions such as calcium, potassium, magnesium and ammonium, already scarce in such conditions, cannot compete successfully for

sites on the cation exchange complex. Failure to find a site means that they are left in the soil solution and are liable to be leached.

Table 5.3 (page 65) illustrates the comparative ability of 'good' and 'poor' humus and the various clay minerals to hold exchangeable cations. This ability is called the cation exchange capacity and is expressed in terms of milliequivalents of cation per 100 g of soil.

Table 8.2 gives some examples of the cation exchange capacity of soils in the British Isles.

Organic matter if well humified has a very high cation

Table 8.2 Cation exchange capacities of mineral soils of different textures, and of organic soils

	Cation exchange capacity in milliequivalents/100 g soil
Coarse sand	2–4
Coarse sandy loam	4–7
Sandy loam	6–18
Loam	10–22
Silty loam	13–28
Clay loam	16–40
Clay	25–50
Well decomposed low-moor peat	180–220
Mor humus	20–50
High-moor peat	30–80

exchange capacity, higher if it is neutral or slightly acid than if it is very acid. Because of this, soils containing 4–8 per cent of good humus have a good capacity for retaining available cations.

The description of cation exchange given here has been greatly simplified. In the field there are continual exchanges from minute to minute involving many different ions. Rainfall will bring an influx of hydrogen ions, from carbonic acid, which will challenge the resident ions. Withdrawal of potassium from the complex by the plant will 'make room' for other ions to move in. Compound fertilizer placed in a narrow band will make the soil solution rich in NH_4^+ and K^+ some of which will exchange with ions which are on the complex. Thus availability of cations on the complex will vary greatly from day to day. Many of the reactions following lime or fertilizer application are cation exchange reactions.

Antagonism is a term to express the competition between different ions for uptake by the plant. It occurs both at the root surface and on the cation exchange complex. If an ion is present in great excess of other ions it will depress the availability of those ions to the plant. The ion present in excess colonizes the cation exchange complex. Other ions are thus driven out into the soil solution and, in our climate, commonly leached. For example, excess calcium will depress the uptake by the plant of magnesium and potassium; excess potassium will depress the uptake of magnesium.

One aspect of antagonism illustrates the need for avoiding excess fertilizer application. High levels of fertilizer potassium can induce magnesium deficiency in crops. In cation exchange terms this can be expressed by:

$$Mg \text{ (clay)} + 2K^+ \rightleftharpoons {}^K_K \text{ (clay)} + Mg^{2+}$$

The Mg^{2+} is then in the soil solution where it will need to compete again with excess potassium for uptake by the plant or it will be leached. As a result of this antagonism, magnesium deficiency is prevalent in greenhouse tomato crops and bizarre examples of potassium-induced magnesium deficiency can be seen in potato crops in which every plant is affected.

Similar equations may be written to express the antagonism of calcium for other ions where excess lime has been applied:

$$K \text{ (clay)} + Ca^{2+} \rightleftharpoons Ca \text{ (clay)} + 2K^+$$

The equation on page 107 illustrates hydrogen/calcium antagonism during soil acidification and results in calcium leaching.

Soil pH, especially if alkaline or strongly acid greatly affects the availability of nutrients. The effect of pH results partly from antagonism as previously described, calcium being dominant in alkaline soils, hydrogen and aluminium in acid soils. Figure 8.2 shows the qualitative effects of soil pH upon the availability of essential nutrients. All elements except manganese and molybdenum have maximum availability in the middle of the pH range (5.5–6.5). Phosphate, potassium and all the trace elements except molybdenum become less available to the plant at high pH values. Phosphate, potassium, calcium, magnesium, copper and zinc also show reduced availability at low pH values. This is partly explained by leaching losses and the antagonism of hydrogen and aluminium ions. The exception to this is phosphate which is not leached but is fixed as low-solubility iron and aluminium phosphates.

Fixation in forms unavailable to the plant is involved in the effects soil pH on phosphate (at both high and low pH values), manganese and boron (at high pH values) and molybdenum at low pH values.

Mineralization of organic matter Organic matter contains the essential elements nitrogen, phosphorus and sulphur in complex forms which cannot be taken up by the plant. These compounds are attacked by micro-organisms and as a result the nitrogenous compounds are converted to ammonium or nitrate forms, phosphorus compounds to available phosphates and sulphur compounds to sulphate.

Some trace elements are also mineralized in this way. The rate at which mineralization takes place depends upon the types of organic substances, the soil pH, the temperature and the presence of adequate supplies of oxygen, and varies from season to season. Obviously, in a season when mineralization is vigorous less nitrogen, phosphorus and sulphur need to be added in fertilizers, but unfortunately the amount released is very difficult to predict.

Physical factors

Root penetration and ramification are important factors in the positional availability of nutrients to the plant root. Soil with good structure encourages the proliferation of tiny feeding roots because it provides adequate oxygen and available water. It is essential that plant roots should not be impeded in any way either by difficulties in penetrating into structural units or by pans. Ions in the soil solution move towards the root zone simply because the water in which they are dissolved is moving. This process is known as mass flow and the plant receives a considerable proportion of its nutrients in this way. Mass flow is obviously prevented by an obstruction, such as a pan, which prevents the passage of water.

If movement of water is prevented ions can still move

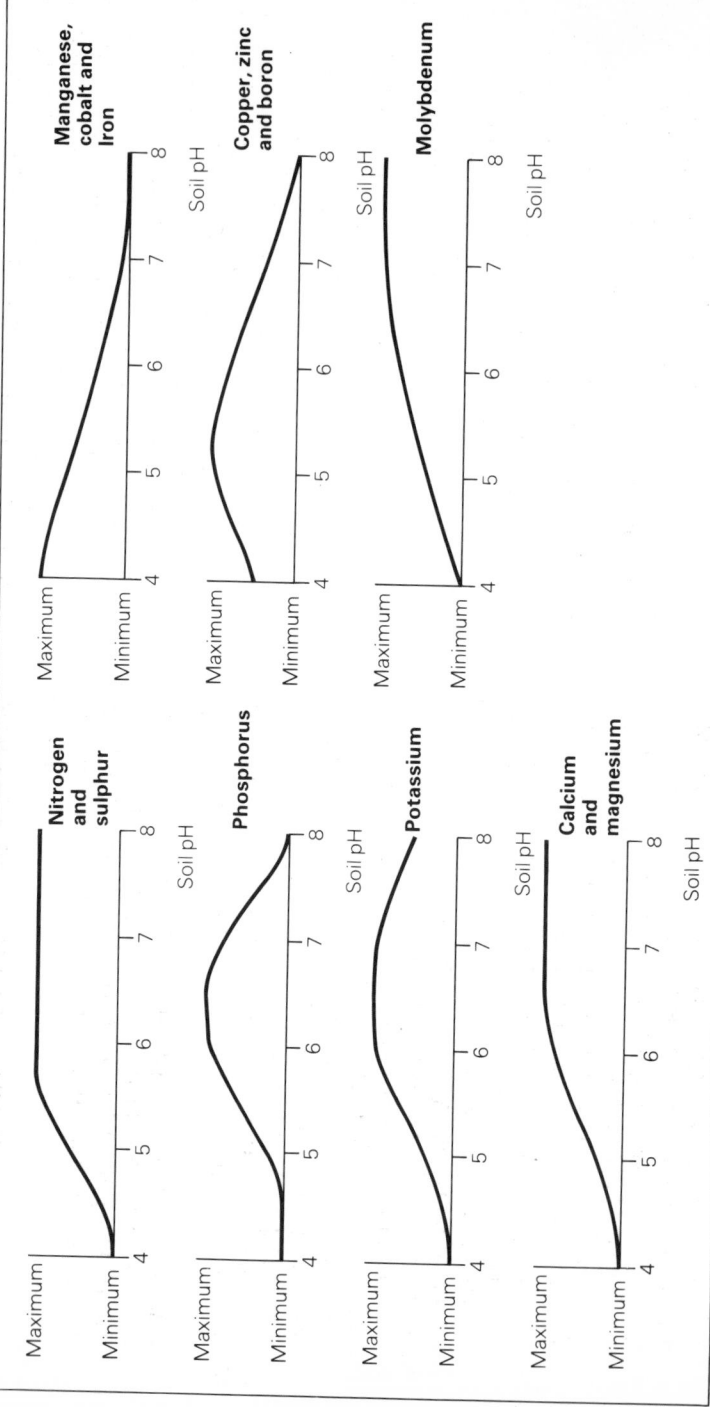

Figure 8.2 Effects of soil pH on the availability of nutrients

within the soil solution by the process of diffusion. In this process the ion moves from high concentration to low concentration. The root, by withdrawing ions from the soil reduces the concentration immediately around it. Provided that there is a continuous film of water, ions will then move from other parts of the soil, where their concentration is higher, towards the root. Some ions, such as potassium, diffuse rapidly through soil water. In this case even a cultivation pan will not prevent the roots from receiving potassium by diffusion. Other ions, such as phosphate, diffuse very slowly and cannot overcome a physical barrier in the same way. Diffusion if ions cannot occur once the soil becomes dry. This is one of the reasons why availability of nutrients depends upon an adequate supply of available water.

Proliferation and penetration of plant roots is very much a function of plant species. Perennial grasses provide the most remarkable example of this as it affects the ability of plants to absorb nutrients from the surface soil. They can take in very much more of a nutrient from soil of a particular availability status, assessed chemically, than any other group of crop plants.

Although the general pathways for nutrients in soils apply to all nutrients each element has its own series of pathways towards or away from the plant root.

Availability of individual nutrients

Nitrogen

Figure 8.3 shows the pathways of atmospheric, soil, plant and animal nitrogen for a grass/clover ley in the absence of fertilizer nitrogen (Fig. 8.3(a)) and with fertilizer nitrogen levels aimed to give maximum yields of dry matter from the sward (Fig. 8.3(b)).

Nitrogen enters the soil in several ways. The main route unless high fertilizer levels are used is through plant residues either directly or after passage through the animal (dung or urine). This nitrogen is in the form of proteins and amino-acids and must be converted by bacteria to ammonium (NH_4^+) or nitrate (NO_3^-) before it becomes available to the plant and is converted back to protein. The stages of conversion of organic nitrogen to ammonium and then to nitrate are carried out by different groups of bacteria. The whole process is known as 'nitrification'.

An important source of soil nitrogen where leguminous crops such as peas, beans, lupins or clover are grown, is that synthesized from atmospheric nitrogen, by bacteria growing in the root nodules, and subsequently released into the soil. A few non-leguminous species also have these nodules. If grass leys rich in clover are used this contribution to available soil nitrogen can be as much as 100–200 kg of nitrogen per hectare each year. The only other 'natural' contribution to soil nitrogen is through rain

(a) No fertilizer nitrogen

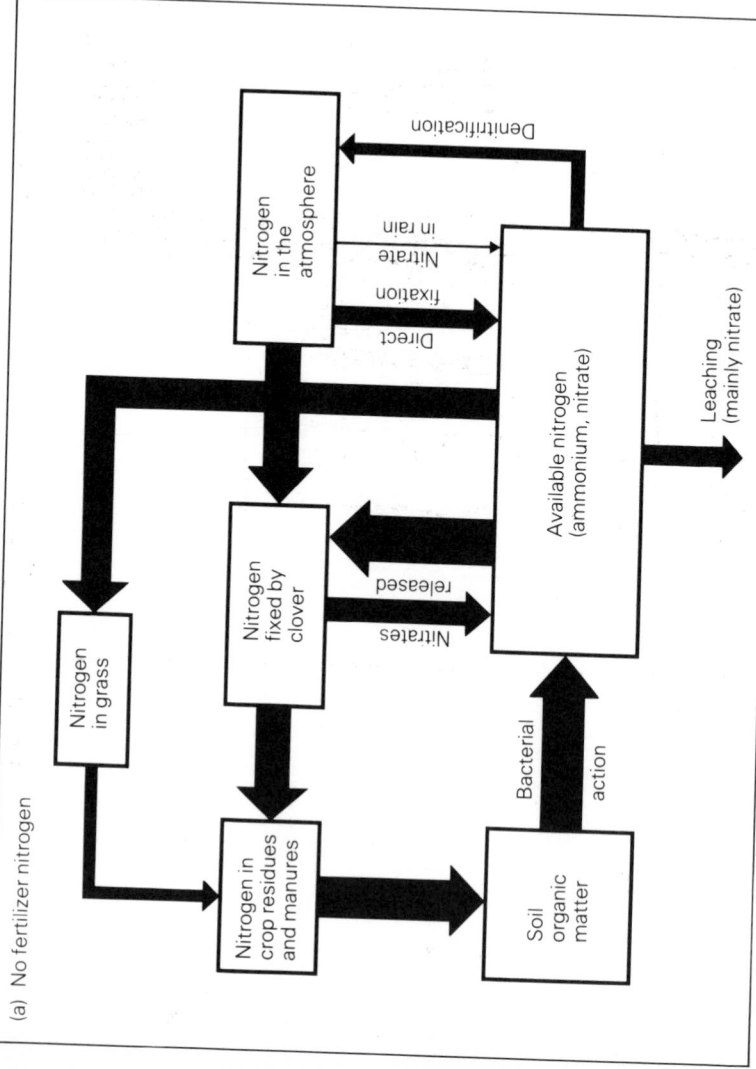

Fig. 8.3 (a) and (b) Pathways of nitrogen in a soil under grass/clover ley

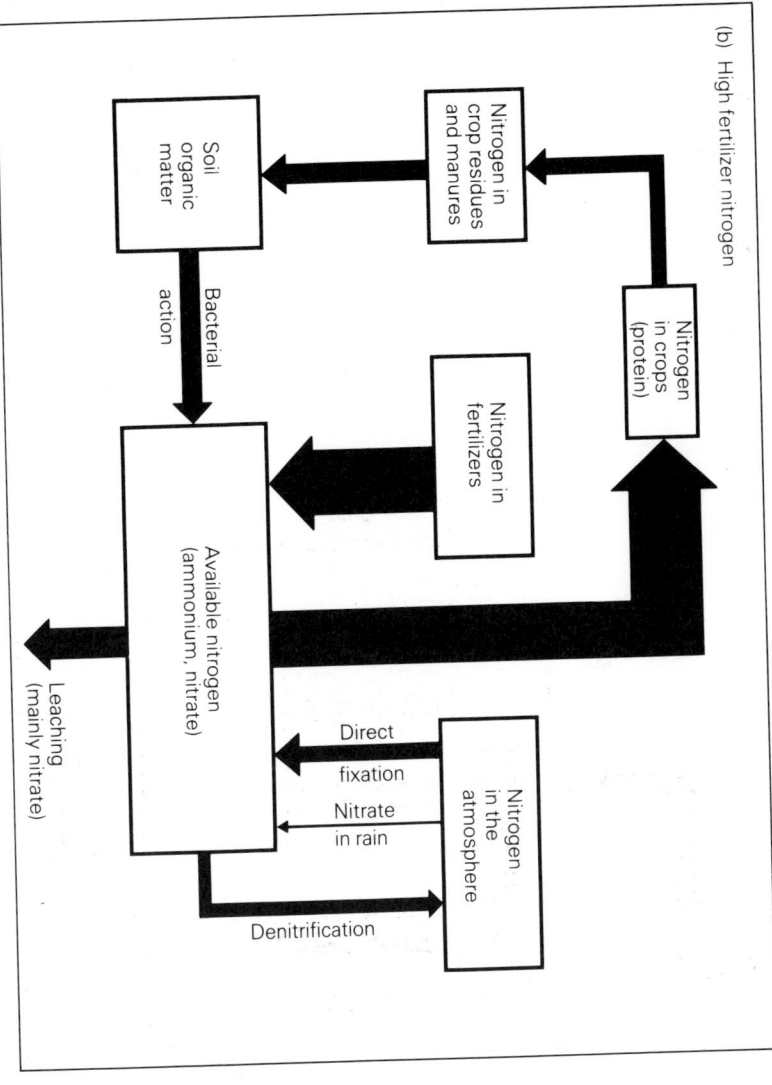

(b) High fertilizer nitrogen

Fig. 8.3 (b)

or snow containing oxides of nitrogen formed during electrical storms. Quite large amounts are involved in some tropical areas but only a few kilograms per hectare are received in this way by soils in the British Isles.

Nitrogenous fertilizers have contributed steadily increasing amounts of nitrogen to the cycle in the last 50 years, giving very large increases in crop production. The most common form of nitrogenous fertilizer is ammonium nitrate, NH_4NO_3, which is included in many compound fertilizers. Both the ammonium and the nitrate are immediately available to the plant, although they can be transformed into less available forms fairly rapidly. Another major source of fertilizer nitrogen is urea, $CO(NH_2)_2$, which is rapidly converted in the soil to NH_4^+ ions. There are also some 'slow release' synthetic materials such as urea formaldehyde and isobutylidene di-urea but they are not as yet important in British agriculture.

Organic manures such as farmyard manure or slurry are another source of soil nitrogen, containing much of their nitrogen in complex forms which are converted to NH_4^+ and NO_3^- by bacterial action.

Temporary storage of nitrogen occurs in the bodies of living soil organisms, the organic matter fraction of the soil and on the clay–humus complex in the form of exchangeable NH_4^+. The living organisms need either to excrete nitrogenous compounds or die before their nitrogen can become available. Some of the nitrogen in organic matter can be fairly readily mineralized to NH_4^+ or NO_3^- but, particularly in acid or peaty soils, much of it is intractable.

The main loss of nitrogen from the soil is through leaching in water passing to the drains. Ammonium nitrogen (NH_4^+) is retained as a cation, although some may be leached if a great excess of another ion is present, but by far the greater part of loss by leaching occurs as nitrate (NO_3^-) which the soil cannot hold against leaching. Appreciable losses of nitrate can occur during short periods of heavy rain, especially from light-textured soils with no crop cover. Ammonium nitrogen is converted by bacteria to nitrate so that it may also be lost by this route.

Nitrogen is also 'lost' in crops. Some of this nitrogen will be returned to the soil in manures but in systems of agriculture such as continuous cash cropping large amounts of nitrogen are removed from the cycle in the soil on which the crops are grown.

Other losses occur by volatilization of ammonia gas (NH_3) from the soil surface but are usually small except where ammonium fertilizers are applied to the surface of Calcareous soils. A much more substantial and less understood loss of soil nitrogen results from denitrification caused by the conversion by bacteria of nitrate to gaseous nitrogen (N_2) which is unavailable to the plant. These bacteria are most active in temporarily or permanently waterlogged soils, especially if they are acid.

Figure 8.3(a) and (b) illustrates important effects on the nitrogen economy of the soil. The comparative length and size of the arrows from one part of the diagrams to another illustrate in general terms the effects of fertilizer nitrogen.

The fertilizer increases the amount of nitrogen in the soil both directly and by increasing the yield of non-harvested parts of the crop such as roots and dead leaves. On the debit side, clover growth is suppressed and the contribution of 100–200 kg of nitrogen per hectare from its root nodules is reduced or lost completely. As a general rule, if more than 150 kg of fertilizer nitrogen per annum are applied there is no point in including clover in the seeds mixture.

Leaching of nitrate nitrogen may also be slightly increased by using nitrogenous fertilizer on the grass/clover crop but the vigorous root system will prevent most of the leaching loss. Much more serious nitrate leaching can occur on row crops especially if all the nitrogenous fertilizer is applied at sowing. Any nitrate leaching loss contributes to inefficient use of nitrogenous fertilizer but, more important, the leached nitrate pollutes watercourses and lakes. In extreme cases this can help to cause rapid multiplication of algae and other organisms in the lake, creating an impossible demand for oxygen. Fish and other creatures are thus asphyxiated.

Phosphorus

Figure 8.4 gives a general picture of soil phosphorus availability. Phosphorus is available to the plant as the phosphate ions HPO_4^{2-} and $H_2PO_4^-$. These are anions and phosphate takes no part in cation exchange reactions.

The plant faces much greater competition for phosphate ions than for available nitrogen. The main reason for this is the fixation of phosphate by the soil. In contrast to nitrogen, very little phosphate is leached from mineral soil in the British Isles. This is also a result of fixation.

The term fixation has fallen into some disrepute among scientists but it is, none the less, very useful and graphically describes what happens. Fixation covers the chemical processes in soil which render phosphate unavailable to the plant, either on a short- or long-term basis. The processes are very complex and have attracted vast amounts of research but the basic position can be summarized quite briefly.

A large proportion of soil phosphate, much more than 90 per cent is, at any time, unavailable to the plant. A further proportion is 'available with difficulty' and a very small proportion, usually less than 1 per cent is available with relative ease. In British soils there is little natural phosphate derived from indigenous rock minerals. Much of the pool of phosphate is derived from the residues of previous applications of phosphate fertilizer, the raw material for which is imported, or from organic manures. Because of the rapid fixation of applied fertilizer phosphate only a small proportion, normally less than 20 per cent is taken up by the first crop after application. Uptake from subsequent crops is very low, commonly between 0 and 5 per cent and, considering also that little or no phosphate is leached, there is an inevitable build-up of *total* soil phosphate wherever phosphate fertilizers are applied regularly.

There is very little water-soluble phosphate even in well-fertilized mineral soils. Any water-soluble fertilizer added, such as superphosphate or mono- and di-ammonium phosphates, is rendered insoluble in the course of days, sometimes hours. It does, however, remain in reasonably available forms for some weeks or months with its availability gradually declining. The rate of fixation depends very much on soil conditions and particularly on pH. The most serious fixation occurs in very acid soil (pH less than 5.0) in which there are large quantities of soluble iron and aluminium hydroxides. These combine with the phosphate ions to give iron and aluminium phosphates which have very low solubility, particularly after they have 'aged' in the soil and have become crystalline. This is a gross oversimplification of phosphate fixation in acid soil but does give the essence of it. This type of fixation declines as pH increases because of the reduced amount of iron and aluminium hydroxides in solution.

In areas of Calcareous soils, where soil pH values exceed 7.0, a different type of fixation occurs as a result of the reaction of soluble phosphates with calcium to give insoluble phosphates such as tricalcium phosphate, $Ca_3(PO_4)_2$. This will occur in fundamentally acid soil which has been limed to pH values above 6.5 but will never be so serious as in soil derived from chalk or soft limestones such as Oolite.

In the middle range of soil pH, 5.5–6.5, fixation by calcium, iron and aluminium is at a minimum.

Phosphate is also 'fixed' by silicate clay minerals such as illite and montmorillonite. Some of this phosphate as well as that fixed directly by iron and aluminium can become available to the plant by a process known as anion exchange in which an anion such as hydroxyl (OH^-) or fluoride (F^-) replaces the phosphate on the clay and thus releases it.

Other agents which compete with the plant for available phosphate are soil micro-organisms which can take up much soluble fertilizer phosphate in the first weeks after application. Such phosphate will be immobilized for some time and will be released only when the organisms die and are mineralized. Soil organic matter contains appreciable amounts of phosphate some of which is in very complex forms. This part of the reserve can be less intractable than the mineral reserve but its release by mineralization is difficult to predict and to define quantitatively. The

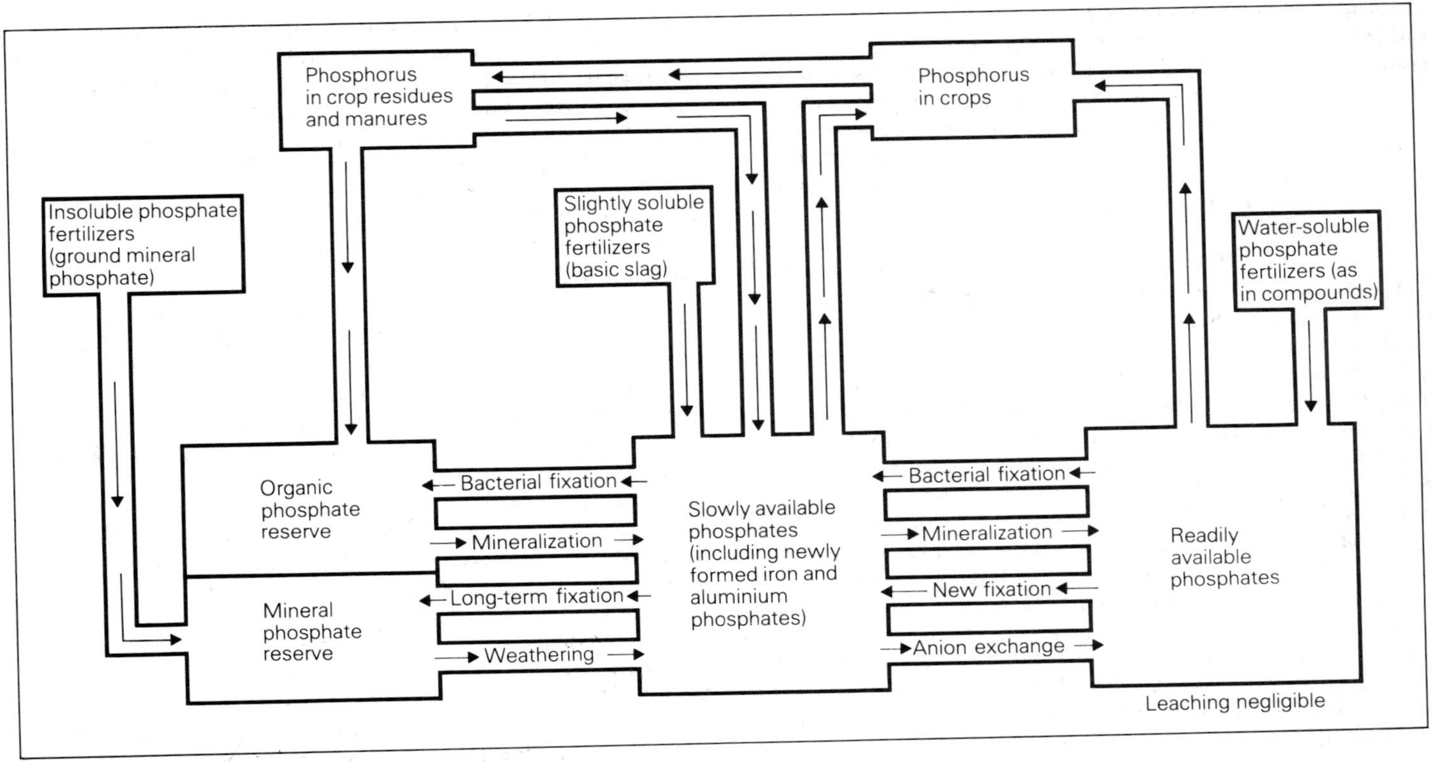

Figure 8.4 Pathways of soil and plant phosphorus

problem lies, not with organic phosphate, newly added as dung or crop residues, but in the large less mobile reserve some of which has been in the soil for years or sometimes centuries. There is evidence that some of this phosphate can be released in the conditions of warm but drought-free summers. This will reduce the requirement for fertilizer phosphate but the amount released cannot be predicted with any accuracy.

Potassium

There is a great reserve of potassium in the soils of many areas of the British Isles. The reason for this is that parent materials, particularly igneous and some metamorphic rocks, contain large quantities of potassium-rich materials such as black mica. For example some schistose rocks of north-west Scotland contain more than 10 per cent of potassium. Some transported parent materials such as glacial till and estuarine alluvium can also contain quite large amounts of potassium. Major exceptions are some sandstones, such as the Bunter Sandstone in Nottinghamshire, some fluvio-glacial sands and gravels, deep peats. Calcareous soils derived from chalk or limestone and coastal sand dunes. All of these can give rise to fundamentally potassium-deficient soils upon which crops will respond vigorously to potassium fertilizer. On other soil types problems will arise only if the rate of

weathering of rock mineral potassium into available forms is too slow to supply crops or if leaching of potassium is high. This can happen on soil derived from potassium-rich but coarsely grained rocks and even on other potassium-rich soils where the demand for potassium by the crop is enhanced by heavy dressings of nitrogenous fertilizer.

The pathways followed by soil potassium are relatively simple as compared with nitrogen and phosphorus. They are shown in Fig. 8.5.

Potassium ions are very 'mobile', moving freely about in the soil solution and even in the plant. They are easily and rapidly adsorbed by the cation exchange complex and are taken up by the plant very readily from either the soil solution or from the cation exchange complex – sometimes preventing the uptake of less mobile ions such as magnesium.

Because of the size of potassium ions, some clays have the capacity to 'fix' and render them temporarily unavailable to the plant by trapping them between the plate-like clay units. This phenomenon occurs mainly in illitic clays which are common in the British Isles. The fixation tends to occurs as soil dries in the summer and some of the potassium is released during the winter as a result of the alternate freezing and thawing of the soil.

Fertilizer potassium is leached readily from light-textured soils low in organic matter but is retained well by other soils unless some antagonistic element – usually

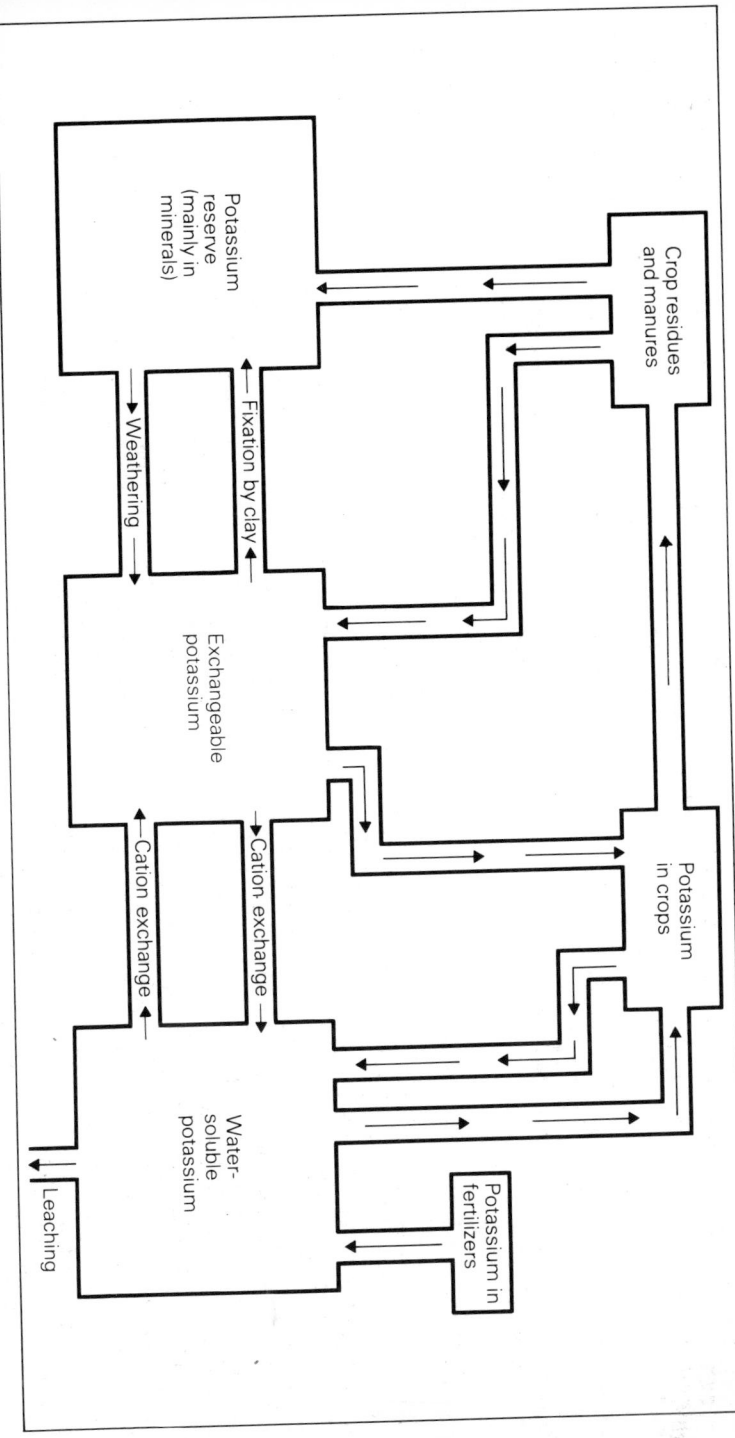

Figure 8.5 Pathways of soil and plant potassium

calcium from liming material – is present in excess. In this case, potassium will be ejected from the cation exchange complex and become susceptible to leaching.

Calcium and magnesium

These elements exist in ionic form on the cation exchange complex. Calcium is the dominant exchangeable ion in most calcareous and mildly acid soils. There are vast reserves of calcium in soils derived from chalk or limestone and of both calcium and magnesium in soils derived from magnesian limestone.

Soils derived from igneous and metamorphic rocks have also considerable reserves of calcium and magnesium because of the wide range, in these rocks, of primary and secondary minerals which contain them.

Availability diagrams for calcium and magnesium are similar to that shown for potassium in Fig. 8.5, except that they are not fixed by clay minerals. Also there is a small organic component of the cycle for magnesium and calcium as both, in contrast to potassium, occur in organic compounds in the plant.

There is very little water-soluble calcium or magnesium in soils. Even when they are added to the soil as ground limestones the solubility is low. Thus, the exchangeable form is by far the greatest source of available calcium and magnesium.

As acidity develops in soils, calcium and magnesium are leached and reach deficiency levels. Correction of acidity by the use of limestones high in calcium and low in magnesium can exacerbate the deficiency of magnesium.

Sulphur

Like nitrogen and phosphorus, sulphur is intimately involved in organic matter, both living and dead. More than 90 per cent of soil sulphur occurs in soil organic matter. It is an essential component of both plant and animal amino-acids and is taken up by the plant in quantities similar to phosphorus.

Most of the sulphur contained in rocks is in the form of sulphides of iron, nickel and other metals. During weathering they are oxidized to sulphates and this is the main form taken up by plants. Organic sulphur compounds must also be mineralized to the sulphate form (SO_4^{2-}) by micro-organisms before they become available to the plant. This occurs most readily in calcareous or well-limed soils at relatively high temperatures (20 °C).

If the soil becomes anaerobic, sulphate is reduced to sulphide (S^{2-}) and the noxious gas hydrogen sulphide (H_2S) is released into the soil. It is toxic to plants.

Carbon, hydrogen, oxygen

These major elements must be regarded separately from the others. They enter the plant mainly in non-ionic form as carbon dioxide, water and oxygen. Provided that a plant is growing well with adequate supplies of other essential nutrients and adequate light, the intake of these elements by the plant will occur automatically.

Trace elements essential to plants

Boron (*B*) is present in the rock mineral tourmaline from which it is weathered as borate (BO_3^{3-}) which is available to plants. It accumulates in soil organic matter and both organic and mineral forms supply boron to the plant. It becomes unavailable in alkaline soils, particularly in seasons dry enough to prevent the mineralization of organically bound boron.

Manganese (*Mn*) is abundant in British mineral soils. It is weathered from rock minerals as the manganous ion (Mn^{2+}) which is available to the plant but can be converted by bacterial action to the manganic (Mn^{3+}) form which is unavailable to the plant. This occurs most readily in alkaline or slightly acid freely drained soils. Very acid soils may contain large amounts of soluble manganese compounds, which are toxic to plants.

Copper (*Cu*) is released from rock minerals during weathering as Cu^{2+} and is adsorbed by the cation exchange complex from which it is available to the plant. Copper also combines with organic compounds to form unavailable substances. Copper availability decreases with increasing pH, but there are also soils (Podsols, Peats) which contain very little total copper.

Iron (*Fe*) is one of the most abundant elements in soils and is available to the plant as the ferrous ion (Fe^{2+}). Like manganese it is oxidized by bacteria in well-drained soils of high pH. The oxidized form, ferric (Fe^{3+}), is unavailable to the plant. There are large quantities of soluble iron in acid soils which, along with aluminium, fix phosphate in forms unavailable to the plant.

Molybdenum is probably taken up by the plant as the molybdate ion (MoO_4^{2-}). In acid soils it is rendered unavailable largely by reaction with iron compounds by which it is tightly held. This reaction is less vigorous at higher pH values and there is little molybdenum deficiency in agricultural soils. There are some areas where parent materials high in molybdenum occur. This brings about molybdenum toxicity in stock grazing on the soils.

Chlorine (*Cl*) is available to the plant as the chloride ion

(Cl⁻). The chances of deficiency in the British Isles are negligible because of chloride in rainwater and fertilizers.

Cobalt in soils is usually associated with manganese oxides, on the surfaces of which it is adsorbed. Its availability to the plant is reduced by the conversion of manganous (Mn^{2+}) compounds to manganic (Mn^{3+}) forms. It is, therefore, least available at high pH values and in well-drained soils.

Physical problems

Part two

Diagnosis of physical soil problems

9

Problems of erosion, drainage, and cultivation-induced pans, involving the physical properties of soil may be anticipated by the use of basic soil survey maps and special interpretive maps (Chapter 4) and thereby prevented. Often, however, the farmer or adviser is faced directly with a problem that has already developed. The detailed examination of the soil profile in the field is a very useful means of diagnosing the causes of physical problems. It can also reveal unsuspected causes for such vexing problems as 'generally unthrifty crops'.

Tools

The tools required are simple and cheap. They are a strong spade, preferably one which has had a triangular lower part of each side of the blade removed to leave a point, occasionally a pickaxe, a strong bladed penknife preferably with one sharp ended blade and one blunt ended (screwdriver shaped) blade, steel measuring tape, and a bucket for baling out water from holes dug in poorly drained soils.

Choice of sites

It is seldom satisfactory, because of soil variation, to select a site for only one profile pit in an area to be investigated.

The aim should be to choose as representative a site as possible and this is not easy. If there is a specific cropping problem, comparison should be made of profile pits in 'good' and 'poor' parts of the field. In other cases guidance can be taken from basic soil maps, topography, the nature of the topsoil – for example its colour, texture and structure, but there is no hard and fast rule. Gateways, areas around food and water troughs and other abnormally treated small areas should be avoided unless you wish to see precisely what poaching and compaction can do to the soil structure. The variability between different parts of a field is often quite astounding especially in some glaciated areas, so several pits should be dug in most cases.

If the soil is under arable crops or temporary grass, it can be useful to dig one profile pit on a nearby site, if available, which has been totally or relatively undisturbed for many years. This may give some clues about the original state of the topsoil before cultivation.

The vegetation and soil surface should be examined before digging the profile pit.

Vegetation

Due to the use of herbicides, there are few reliable clues in the weed flora of cultivated soil or temporary short-term grassland to the nature of the soil. In pasture, the state of the vegetation gives obvious clues to poaching caused by overstocking. In relatively undisturbed hill and marginal areas the vegetation is a much better guide to what lies beneath as there are very close soil/natural vegetation relationships in undisturbed soils.

Soil surface

Wherever possible the soil surface should be examined when there is no crop cover. It is useful to do this two or three times; immediately after ploughing, after frost and rain and after secondary cultivations.

The general colour impression of the dry soil can be useful. Greyish colours indicate possible poor surface drainage; dark colours, peatiness; and brownish colours, good surface drainage.

If recently ploughed, the way in which the furrow slice stands or spreads is an indication of texture, clay and clay loam standing stiff and often setting hard on drying out whereas sands and silts will form gentler ridges. Crumbliness of the furrow slice after frost should also be noted.

If secondary cultivations have been made, the presence and thickness of surface caps or crusts is easily seen. The caps can be peeled off by using a sharp blade and some assessment can be made of the toughness of the caps.

Signs of standing water or the glistening appearance
associated with very wet soil should be noted.

Signs of worms and moles indicate that the soil is not
acutely acid and is reasonably well drained.

The profile pit

Digging the profile pit

The pit should be dug sufficiently large to allow detailed
examination of one face without being cramped. The
operator needs easy access to the lower parts of the
profile. Ideally the pit should be about 1 m × 2 m and it is
usually sufficient to dig to a depth of 1 m.

It is essential to dig the profile pit yourself. There is
much to be learned during the digging, particularly about
the relative hardness and compactness of different
horizons.

If the spade meets with resistance check whether it is
caused by stones or by a pan. The shape, size range and
variety of origins of stones should be noted in order to
assess the type of parent material.

Cleaving of the soil to the spade gives an indication of
clay or clay loam texture.

Comparative wetness of different horizons gives a first

indication of restricted water movement.

Colour, texture and structure changes identify horizons.

Preparation of the face of the pit

After digging, one of the shorter sides of the pit should be
'dressed up'. The spade will have destroyed structure at
the face. Colours will be smeared together and lines of
natural seepage of water sealed. If the pit is situated on a
slope, the long axis should run up and down the slope and
the short upslope side should be selected for early
observations of water seepage points. In extreme cases it
may be necessary then to resort to the downslope face to
avoid blurring by the seeping water. The aims of cleaning
up the face should be to expose as nearly as possible the
original structure, colours, worm channels and horizon
differences of the soil. This can be done with the aid of a
sharp blade, by probing gently to find natural cleavages
and channels in the soil. It is very helpful in some types of
soil to leave the pit, after digging, for two or three days to
allow the smeared surface to dry a little, after which it is
easier to identify structure cracks. Leached pale grey
horizons in Podsols are also much more easily seen after a
lapse of two or three days. In poorly drained soil the
water-table will also probably be evident after this period.
If there is strong evidence of greyness due to gleying it is

better to examine the lower horizons as soon as the pit is
dug, before it fills up with water!

General inspection of the face

Note the number and depth of horizons identifiable by
sight or by 'feel', including the cultivated layer as one
horizon unless a compacted zone is found within it. In
cultivated soils particular attention should be paid to the
bottom of the ploughed layer and the zone immediately
below where pans may occur.

Colour changes down the profile should be observed.
The zone of maximum greyness and the type of mottling
(around root channels or along structure cracks) in a Gley
soil will help to decide whether you are dealing with a
Surface-Water Gley or a Ground-Water Gley. Darkness of
colour is a pointer to the organic matter content.

The pattern of root penetration and ramification is most
important in crop problems and gives excellent indications,
during the growing season, of pans and compaction. Points
to note are signs of horizontal rooting or deformed roots in
tap-rooted crops, frequency of fine roots with depth,
rooting within or between structure units, a steady
reduction in roots with depth or a sharp lower boundary.
Depth and frequency of worm channels indicate organic
matter incorporation and the position of the water-table.

Points of vigorous water seepage into the pit, preferably
a day or so after heavy rain indicate the upper surfaces of
pans. Position of the water-table after leaving the pit open
for two or three days will be of some help in determining
the drainage status but should not be relied upon as the
sole indicator because it is dependent upon recent rainfall
and speed of movement of water through the soil into the
pit.

Inspection of individual horizons

Assess the texture by the method described in Chapter 6.

- Measure the depth of each horizon.
- Note the nature of the boundaries between horizons.
  Sharp boundaries indicate a very abrupt change of
  properties, either natural, as in the very sharp change
  from mor humus to the mineral horizons in the Podsol
  or man-induced as at the lower boundary of many
  cultivated soils.
- Assess the structure. This is not easy and comes only
  with a good deal of practice. The beginner should not
  be greatly discouraged by this as many conclusions that
  can be made from a knowledge of soil structure can be
  inferred from other observations. One well-respected
  soil surveyor was in the habit of assessing soil structure
  by taking a spadeful of soil, dropping it on a hard

surface and observing how it broke up. Some types of structure are more easily identified than others and strongly developed units are very obvious.

Crumb and granular structures are usually associated with intense root ramification and 'hanging' from the profile around roots. Massive and prismatic structures usually occur in subsoil especially with poor drainage but intermediate blocky structure types and more difficult to identify.

Angular blocky structures in heavy soil often show up as the face dries out. Sub-angular blocky structures are more difficult to identify but may be found immediately below horizons of crumb or granular structure. Platy structure can be very obvious in the case of some cultivation pans but in other cases requires careful profile inspection to identify the horizontal orientation of structure units.

• Assess the organic matter content and type – purely organic horizons are easily identified. In undisturbed soil, the nature of the organic horizons is very important and assessment can be made of the build-up of undecomposed, partially humified and wholly humified organic horizons in Peats and Podsols. It is much more difficult to assess the amount of incorporated organic matter in cultivated soils although it is usually related to darkness of colour. Confirmation by laboratory analysis may be required if this property is of particular interest, as in the assessment of erosion risks.

• Assess the colour. This is a very important characteristic and the colour-blind person will be at a disadvantage as the colour differences in soils can be quite subtle. The main use of soil ucolour is in assessing the drainage status using the blue grey/orange brown mottling as an indicator. It must be stressed that absolute reliance should not be put on colour as a diagnostic tool. For instance in soils derived from strongly coloured parent materials such as Old Red Sandstone the rich red-brown inherited colours are very persistent and can mask the normal gley mottling.

Greyness caused by leaching, in the Podsol could be confused by the beginner with greyness caused by waterlogging in the Gley but consideration of other soil properties should correct this misapprehension.

The most important drawback to the use of colour in assessment of drainage conditions is the time taken for colours to change after drainage has been artificially improved – gley colours can persist for many years after the factors causing them have been removed. Thus, taken alone, without reference to drainage history and water-table position measurements, misleading conclusions can be made.

• Confirm the presence and nature of pans. Evidence will already be available from the resistance to the spade during digging, the percolation of water into the pit and the root pattern. The presence of a pan can sometimes be confirmed visually as is the case with iron pans. In other cases the extent and toughness of the pan may be confirmed by pushing a strong knife blade into the profile, *both horizontally and vertically*, very frequently and assessing the resistance subjectively. The 'knack' of doing this is readily acquired. Vertical probing is essential for pans with platy structure as resistance to the blade will be much greater vertically than horizontally. In order to probe vertically, the face of the pit should be cut back a few centimetres at a time to give a series of horizontal surfaces.

The cultivated layer

In cultivated soils, the subsoil horizons yield information about the genesis of the soil, its inherent properties such as its natural drainage and its basic cropping potential.

The cultivated layer and the effects of cultivating implements on the layers immediately below it must be viewed differently. Natural structure and other properties have been modified, sometimes very greatly, by liming, cultivations, manuring and fertilizer application. In these zones the size and distribution of clods within the cultivated layer should be observed and related to rooting potential and emergence problems.

Pans may be found within the cultivated layer or, more frequently, immediately below it where they may be 1–10 cm thick. These cultivation pans can occur in soil of all texture types.

Horizontal smearing and polishing can occur at the lower boundary of the cultivated layer cutting off drainage channels. This can sometimes be identified by water seepage but, more reliably, by a smooth shiny surface below the plough, often bearing the mark of the share.

Anaerobic pockets, where straw or grass turf has been ploughed in under very wet conditions, can be identified by their dark grey colour and the smell of hydrogen sulphide.

Deductions from soil profile examination

It is very unwise to use in isolation any one of the properties of the profile to diagnose the potential of the soil for cropping and the problems that might be experienced. But, all the observations and properties taken together provide a very strong basis for diagnosing the causes of cropping problems.

Several aspects of diagnosis from field observations can be supported and confirmed by laboratory tests or by the use of more sophisticated equipment in the field.

Laboratory tests

Organic matter content may be measured approximately by determining the loss in weight of dry soil when it is ignited at about 500 °C. The organic matter is burnt in this way and the loss in weight is expressed as a percentage of the dry weight of the soil. The measurement helps to assess the likelihood of cultivation, erosion, drought and nutrient availability problems.

Particle-size analysis is a term used for laboratory determinations of texture. Organic matter and calcium carbonate are first removed by chemical agents. Then the coarse and fine sand fractions are determined by wet sieving. The silt and clay fractions are too fine to be estimated in this way and are determined by the rate at which they fall through water, which is related to the size of the particles. The maximum sized silt particle (0.02 mm) will fall 100 times as quickly as the maximum sized clay particle (0.002 mm).

Using particle analysis the place of the soil on the texture triangle (Fig. 6.1) can be established. It serves to confirm texture assessment made in the field.

Water content may be measured by drying a representative sample taken from the field at 105 °C for three hours. It is used mainly in assessment of irrigation needs and, in conjunction with dry bulk density measurements to assess aeration (Fig. 7.5, page 99). Its use in estimating irrigation needs is limited because of wide fluctuations both in time and from point to point in the field.

Dry bulk density is measured by using cores of approximately 10 cm diameter at various depths in the profile, using a cylindrical corer. The water content is determined by drying the core and the dry bulk density is then measured as a ratio of the weight of the dry core to that of the same volume of water. This is a most useful measure in confirming the presence of pans and surface compaction and the likelihood of anaerobic conditions.

Plastic limit is measured by rolling wet soil into threads of 3 mm diameter. As rolling continues the soil dries out. The percentage of water in the soil when it just begins to crumble is the plastic limit. This property is important in deciding when a soil is too wet to be cultivated without damage to structure. Light-textured soils with less than 10–15 per cent clay are non-plastic and have therefore no plastic limit.

Structure stability is usually measured by putting the structure units on a sieve and moving the sieve up and down mechanically in water. Aggregates which cannot resist the water movement break down and pass through the sieve. The proportion remaining on the sieve can be dried, weighed and expressed as a percentage of the total weight of the sample. Great care is needed to wet the aggregates very slowly before the test is made, otherwise air escaping explosively from the aggregates can disrupt them.

This is a useful tool for predicting structure breakdown in the field and results of this, such as capping.

Field tests

Resistance to root penetration can be measured in various ways. The method most helpful in the field uses a penetrometer. This is a simple device for giving a quantitative value to the knife-blade probing described earlier in this chapter. Tips of various shapes, usually cones, can be fitted to a rod which is pushed into the soil. As it forces its way into the soil the resistance to its passage is measured using a spring. This device is helpful in confirming the presence, thickness and strength of pans.

Position of the water-table and its fluctuations may be measured by the use of dip-wells or the more refined piezometer. A dip-well is simply an open auger hole (8–12 cm diameter) in the soil, the level of water in which can be measured by a dipstick. Unfortunately the position of the water surface in a dip-well may give a false impression of the ground-water-table because water may percolate in from a pan higher up the profile. The piezometer will prevent this. It is constructed in a dip-well auger hole. First a layer of coarse sand is introduced at the bottom of the hole. A narrow (1 cm diameter) plastic pipe is inserted in the middle of the hole and surrounded by bentonite or other suitable material packed between the pipe and the sides of the hole. The narrow pipe is needed to ensure rapid response of the water in the pipe to water-table changes. If surface water above a pan is suspected piezometers of various depths will establish the position of both surface and ground water.

Water tension is measured by the use of a tensiometer, consisting of a porous ceramic cup connected to a vertical non-porous tube, the whole being filled with water. The top of the tube is connected to a vacuum pressure gauge. The ceramic cup is introduced to the soil and the suction of the water, which varies according to the dryness of the surrounding soil, is recorded on the gauge. Tensiometers are useful only at tensions below 0.85 bar. At this stage sandy soils are depleted of water but soils of heavier texture will still contain available water. Thus, as a guide to irrigation needs tensiometers are useful mainly for plants which require the soil to be kept continually moist (market garden conditions). If the tensiometer readings approach 1 bar the water column will break and air will enter the tube. Re-charging with water is then necessary.

Problems of various texture groups

Many of the soil physical problems described in Chapters 10–13 are associated with particular groups of soil texture which may be described as 'light', 'medium' and 'heavy' textured. These groups bring together several of the texture types shown in Fig. 6.1 (page 74). The broad grouping used in the following chapter is shown on the texture triangle in Fig. 9.1. The divisions are arbitrary and some anomalies will be found. For example, soils of light texture include sands, loamy sands and most sandy loams but if soils of these texture types contain high proportions of fine sand they tend to behave more like medium-textured silty soils.

Medium-textured soils include most sandy clay loams, some fine sandy loams, loams, silt loams and silts. Again, there might be a case for separating the silt loams and silts into a separate category of 'silty soils', but the broad medium-texture grouping will serve for most purposes.

Heavy-textured soils include sandy clays, clay loams, silty clays, silty clay loams and clays.

Problems within each group tend to become more difficult as the soil texture approaches the apexes of the triangle. For example sands will drought more easily than sandy loams; clays will present more cultivation difficulties than clay loams. Table 9.1 gives a general summary of the physical problems associated with soils of the three main texture groups.

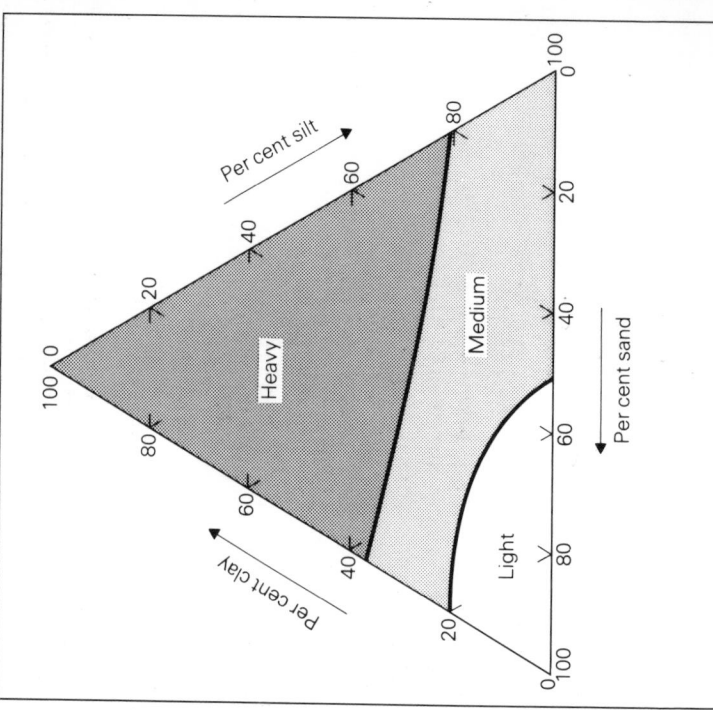

Figure 9.1 A grouping of texture types

Table 9.1 Problems associated with soils of light, medium and heavy texture (see Fig. 9.1)

Texture	Availability of water	Availability of nutrients	Other problems
Light	Drought occurs readily. Drainage is too free. Available water capacity is low. (If there is a high proportion of fine sand, problems of excess water are as for 'Medium')	Past leaching has commonly resulted in low content of many nutrients. Deficiencies of trace elements, e.g. manganese, copper, boron, are easily induced by liming. Applied lime and fertilizers are subject to leaching	Puffy seedbeds occur due to lack of compaction after late ploughing. Capping can occur if the soil is high in fine sand, restricting the emergence of small-seeded crops. Wind erosion occurs if the soil is low in organic matter and high in fine sand. Pans can form despite the light texture if cultivated when too wet. Root penetration can be greatly reduced. Organic matter is easily lost by oxidation and is difficult to build up
Medium	Problems of excess water arise despite high available water capacity if cultivated when too wet, causing pans or surface compaction	Problems are infrequent except in intensive systems of cropping with inadequate input of fertilizers and manures, or where pans prevent root ramification	Serious capping problems can occur if the soil is high in silt, restricting emergence of all crops. Water erosion can occur, especially if the soil is low in organic matter and high in silt. Pans may form if the soil is cultivated when too wet
Heavy	Drainage is commonly restricted and excess water can be found unless underdrainage is efficient	As for 'Medium'	Soil structure is poor except in some calcareous clays. Soils are cold and slow to warm up in the spring. Cultivation difficulties, arising from wetness, give rise to compaction, pans, clod formation and poor seedbeds

Maintenance of organic matter and structure

10

Organic matter content and soil structure are closely linked. Chapters 5 and 6 have illustrated the advantages of improving or maintaining these two properties in terms of water availability, ease of cultivation, nutrient availability and root ramification.

Anything that increases the content of well-humified organic matter helps also to improve structure. The general rules for organic matter conservation and structure improvement are, therefore, similar although some aspects of structure maintenance do not involve organic matter.

Unfortunately it is much easier to destroy organic matter and structure than it is to maintain them, but there is much that the farmer can do, within economic systems of farming, to protect them.

General rules

● Aim for the largest possible return to the soil of crop residues.
● Disturb the soil by cultivations as little as possible.
● Keep crop cover at a maximum.
● Maintain an adequate supply of lime to encourage micro-organism and earthworm activity.
● Ensure that drainage is satisfactory but not excessive.
● Use green manuring crops whenever possible.
● Remember that grass/clover crops are more effective

than arable crops unless maximum returns of straw and other crop residues are made.

- Ensure that bulky green crops are well bruised and crushed and straw is finely chopped before incorporation.

- Incorporate manures and crop residues as quickly as possible to avoid oxidative losses.

Maintenance or build-up of organic matter

If organic matter contents have become very low, it is a slow business to build them up again to adequate levels especially on light-textured soils. It can also be expensive unless it becomes part of the policy in an economic management system and it cannot be stressed too much that good systems of general farm management are important in organic matter conservation.

That the organic matter content of soils in the British Isles can be increased to 10–20 per cent has been amply demonstrated in many country house gardens or market gardens. There is a striking example at a market garden near Edinburgh, which has been in the same family for a century. A small part (2 ha) of this holding, on a raised beach soil of loamy sand texture, has received approximately 250 t/100 ha of farmyard manure, or its equivalent, annually for 100 years.

Judging by nearby farm soil the original organic matter content of the surface soil was probably 2–3 per cent with very little below the plough layer. The present organic matter content, in the top 40 cm, is about 20 per cent; at 120 cm, 8 per cent; and at 180 cm (6 f), 5 per cent. The earthworm population at 180 cm depth is higher than that in many permanent pastures.

This soil has received approximately 6000 t of *dry* organic matter during 100 years and now contains about 2000 t/ha to a depth of 180 cm. Thus about 35 per cent has been retained. Inevitably there will be a large margin of error in these estimates.

Such a soil, which produces excellent market garden crops is virtually invaluable. If mixed and bagged it could be sold as an excellent organic fertilizer.

Less exciting, but better documented evidence from field experiments has demonstrated that between 10 and 40 per cent of added organic matter is eventually retained as humus. The best 'guestimate' for general purposes is that, in north-west Europe 25–30 per cent of added organic matter is retained.

Growers of high-value crops who aim to build up soil organic matter to high levels can do so only by importing large quantities to their holdings. Such methods are quite unrealistic for the usual range of agricultural crops, in terms of both cost and availability. It is therefore unwise, in normal agricultural practice to maintain organic matter above a level consistent with the avoidance of erosion and damage due to cultivations and with good crop yields.

Annual requirements

The amount of organic matter that will need to be added to the soil to maintain the organic matter content will vary a good deal. Annual losses of organic matter from soils under a stable management system in cool humid climates will range between 0.75 and 1.25 t/ha. Any material added to replace these losses will contain water. Also a large proportion of the added organic matter will be lost during decomposition. For example farmyard manure usually contains 20–30 per cent of dry matter. Assuming this and a loss of 75 per cent of the dry matter during humification, some 10–25 t of farmyard manure would need to be applied every year for maintenance. The lower values (10–17.5 t) would apply to heavy-textured soils in cool, wet parts of the country and the higher values (17.5–25 t) to light-textured soils in drier areas.

Figure 10.1 shows the range of annual inputs of dry organic matter from various sources set against the likely losses.

Sources

Farmyard manure Traditional farmyard manure, made with straw will contribute only 0.5–0.8 t of humus for

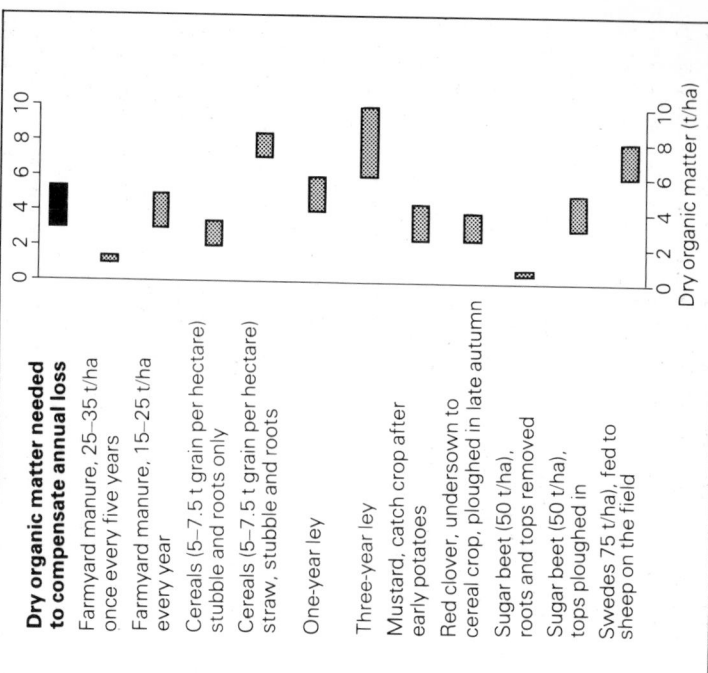

Dry organic matter needed to compensate annual loss

Farmyard manure, 25–35 t/ha once every five years

Farmyard manure, 15–25 t/ha every year

Cereals (5–7.5 t grain per hectare) stubble and roots only

Cereals (5–7.5 t grain per hectare) straw, stubble and roots

One-year ley

Three-year ley

Mustard, catch crop after early potatoes

Red clover, undersown to cereal crop, ploughed in late autumn

Sugar beet (50 t/ha), roots and tops removed

Sugar beet (50 t/ha), tops ploughed in

Swedes 75 t/ha, fed to sheep on the field

Dry organic matter (t/ha)

Figure 10.1 Range of annual gains and losses of organic matter from soils

every 10 t applied. The higher figure refers to well-rotted, well-stored manure which will have a high dry matter content and in which part of the humification process will have occurred in the manure heap. Even where making farmyard manure is an integral part of the farming system it is usually not possible to apply, on a rotational basis, more than 25–35 t/ha once every 4–6 years – equivalent to approximately 4–9 t per annum. This is only one-quarter to one-third of the amount shown by field experiments (15–25 t/ha annually) to be necessary for maintenance.

None the less, well-made farmyard manure makes a useful contribution to organic matter maintenance. It should be applied, if feasible, to bare soil and incorporated immediately after spreading. Application to grassland is relatively inefficient because of oxidative losses, the impossibility of mechanical incorporation and overall contamination of herbage. This is doubly unfortunate because grassland farms produce most farmyard manure and, even if there is an alternative holding on which to use the manure, it must be very close by to justify transporting the material.

Slurry from cattle and pigs is urine and faeces diluted with water used to wash down the living quarters and stored in liquid form. It is very variable in composition but usually has a low carbon/nitrogen ratio compared with farmyard manure which contains a lot of straw. Slurry will decompose readily in soil and will yield less humus per unit of dry matter applied than farmyard manure. The same rules for application should be followed as for farmyard manure, but this is more difficult in the case of slurry because of the problems of disposal of large volumes of liquid at regular intervals.

Green manures are crops grown specially to incorporate into the soil without removing any of the crop for commercial purposes. Green manuring is a very old-established process and was used by the Romans with leguminous crops such as vetches and lupins.

The essential characteristics of green manure crops are rapid growth, vigorous root development and abundant tops. Cheap, reliable and rapidly germinating seed is essential and the crop must not present husbandry problems.

Green manure crops perform several functions:

● Supply of organic matter to the soil.
● Addition of readily mineralized nutrients, particularly nitrogen, for the following crop.
● Provision of soil cover, important in areas where erosion is a problem.
● Prevention of leaching of nutrients by recycling them.
● Control of weeds by smothering them – a function which has declined in importance as weedkillers have become more effective.

Unfortunately the two main functions, building up organic matter and supplying available nutrients, conflict with each other. Green manures, properly used, can *either* increase humus content *or* increase the available nitrogen supply but cannot effectively do both at the same time.

If the aim of growing a green manure crop is to supply available nitrogen a legume should be grown and should be ploughed in while immature. This gives material with a very low carbon/nitrogen ratio and a high proportion of readily decomposable organic compounds. As it releases available nitrogen much of the organic matter is lost as carbon dioxide to the atmosphere so that humus content is not greatly increased. Suitable leguminous crops are red clover (*Trifolium pratense*), sweet clovers (*Melilotus* spp.), common vetch (*Vicia sativa*), trefoil (*Medicago lupulina*) and yellow lupin (*Lupinus luteus*).

If the main aim of growing a green manure crop is to increase organic matter content it is preferable to grow a non-leguminous crop and to allow it to grow further towards maturity when it will have a higher carbon/nitrogen ratio and will contribute persistent humus to the soil. Suitable species are mustard (*Sinapis alba*), rye (*Secale cereale*), rape (*Brassica napus*), turnip rapes (*Brassica rapa* and crosses) and ryegrasses (*Lolium* spp.).

In practice the long-term aims of increasing organic matter often take second place to the short-term benefits of supplying available nitrogen cheaply. Even in these circumstances, green manures in cool humid climates can make contributions to soil organic matter about 60–80 per cent of those made by similar additions of dry matter as farmyard manure.

Some growers, in the days before effective herbicides were available, were willing to sacrifice one whole season out of four or five to green manuring and cultivations to clean out perennial weeds. Such practice could not be economical today and the green manure crop must be fitted in between economic crops. This can be done by sowing immediately after harvesting an early crop or by undersowing a cereal crop with a green crop. The latter method has become more difficult as cereal yields have increased and the undersown crop tends to be smothered. On the other hand, too vigorous an undersown crop can cause harvest problems. It might be worth while using a low-growing legume such as wild white clover (*Trifolium repens*) or bird's foot trefoil (*Lotus corniculatus*) for undersowing.

Fertilizers may be applied to the green crop but there are usually sufficient residual nutrients from the previous crop. Certainly no nitrogenous fertilizer should be applied to a leguminous green crop.

It is important to bruise or chop bulky green crops thoroughly before incorporating them in the soil. Great care should be taken to avoid the formation of a discrete layer of organic matter which would tend to become anaerobic.

Feeding crops on the field   The practice of feeding standing crops on the field is a useful variant of green manuring. A typical example is the controlled feeding of standing crops of swedes or turnips to sheep. Faeces, urine and uneaten parts of the crop are returned to the soil and lightly trodden in during the winter when oxidative losses are at a minimum. The system has management and economic limitations but is certainly an efficient way of returning organic matter to the soil.

The grass crop   Soil under good permanent pasture with its lack of disturbance, constantly renewed root system and its supply of clover nitrogen will always have a much higher organic matter content than adjacent arable land (Table 5.4). Organic matter content declines rapidly after pasture is ploughed out (Fig. 5.2) and reaches a new equilibrium level over several years. It takes a very long time to restore the original level by returning the arable to grass. Cases have been recorded in which 25 years under grass have not nearly restored the organic matter content of old arable to that of adjacent permanent pasture.

Short-term grass, unless left down for three years or more, does not have time to establish the intimate root system of permanent pasture with its consequent large contribution to soil organic matter and those considering changing from a ley/arable system to continuous arable must assess the comparative effects of short-term grass and

arable crops on the organic matter content of the soil. Figure 10.1 gives some guidance. The balance between the two probably depends upon the intensity of production in the grass break. Intensively managed three-year grass where some 12–13 t/ha of dry matter are being harvested will develop a well-ramified root system and will certainly contribute more organic matter than three years of cereals,

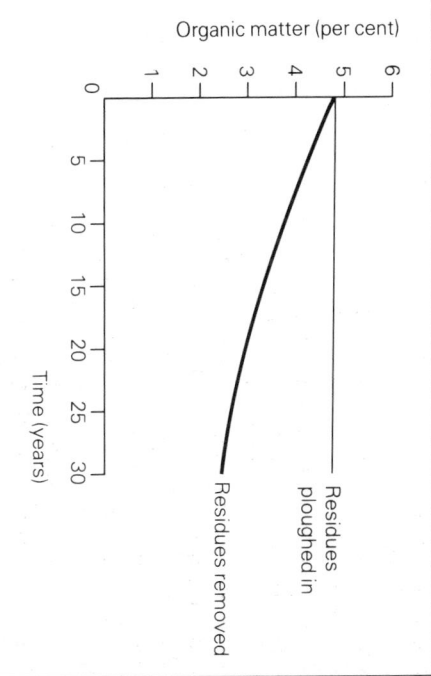

Figure 10.2   Comparison of organic matter content of a soil with and without ploughing in of crop residues

even if straw is returned to the soil. Less intensively managed grass will probably contribute no more than cereal straw plus stubble (5–7.5 t/ha per year).

Arable crop residues The residues from commercial crops contribute considerably towards the maintenance of soil organic matter. It is a good rule to return all the non-economic part of the crop to the soil and to incorporate it as quickly as possible. The production of heavy crops generally means the production of large crop residues and in cool, humid climates increasing yields by means of optimum lime and fertilizer usage may actually maintain the organic matter content of soils if all residues are ploughed in. Figure 10.2 shows an example of this over a period of 30 years.

The special case of straw Very large amounts of straw have been burned, particularly in south-east England over the last 20 years. The fashion comes and goes according to the value of straw, the convenience to the farmer and the disease hazards reputed to be reduced by burning. Ploughed-in straw can make quite large contributions to soil organic matter, particularly where there is continuous cereal production. Even a short strawed barley will yield, not counting normal stubble 2.5–3.5 t/ha of dry straw organic matter. Much of this material should persist in soil. Because of its high carbon/nitrogen ratio (40–80) early

decomposition is helped by applying 1 unit of fertilizer nitrogen for each 100 units of straw ploughed in. Chopping the straw also helps decomposition.

There is no doubt that straw which has been badly incorporated , particularly in poorly drained soil, can lie in a discrete layer and decompose anaerobically giving rise to a noxious smell and possible toxicity to plants. This has led to some prejudice against straw incorporation as has the fact that the farmer will see no immediate return for his effort whatever. However, for the long-term good of soil organic matter all straw which is not used for bedding and feeding should be returned to the soil.

Structure maintenance

In addition to the methods of supplying organic matter described in the first part of this Chapter there are several ways in which structure may be maintained or improved.

Minimum disturbance

There is no doubt that soil undisturbed by cultivation will develop and retain a better structure than a similar cultivated soil. Now that the technology of weed control, especially of pernicious perennial weeds such as couch grass (*Agropyron repens*), has become advanced, one of the main reasons for intensive cultivation has gone.

Direct drilling of seed following overall herbicide application can be applied successfully on many soils containing less than 25 per cent clay and even on some heavier soils provided that the original structure of the surface soil is satisfactory. More often it will be necessary to use some light surface cultivation of the top 5–10 cm to create suitable conditions for sowing seed. This technique is an example of minimal cultivation.

It must be stressed that either technique is very difficult to apply to poorly structured heavy soils in which roots do not ramify easily. Yields equivalent to those under traditional cultivation systems can sometimes be achieved but only by using some extra 50–70 kg/ha fertilizer nitrogen, the cost of which might well outweigh the energy savings made by minimal cultivation.

Drainage and/or subsoiling of poorly drained soils will improve structure by encouraging drying out and opening up cracks.

Timing of cultivations Under traditional systems of cultivation it is important that ploughing should not be done on soils containing more than 20 per cent of clay when the soil is wet enough to be plastic – that is above the plastic limit. If this is done, particularly at the same plough depth each year a plough pan will result having platy or laminar structure. This will restrict the effective depth of the soil to the plough layer and will give rise to surface water problems.

It is particularly important not to use rotary cultivation implements on heavy soils in plastic conditions because the blades at the bottom of their orbit smear and polish the soil and very effectively seal off vertical drainage cracks.

A good general rule is to plough as soon as possible after harvesting, before the early winter rains have brought the soil to field capacity. This has an added advantage of increasing the chance of producing a frost tilth during the winter.

In wet areas of the country, especially in wet seasons, it is just not possible to plough clay soils under sufficiently dry conditions to avoid creating a pan. The problem can be lessened, but seldom solved, by ploughing at different depths each year or by use of a chisel attachment when ploughing to break up the pan. The only satisfactory solution is to accept that heavy soils in wet areas must be used mainly for grassland. Even if this is done incalculable damage to soil structure can result from surface trampling of overstocked pastures.

Avoidance of surface compaction Increasingly heavy machinery and implements have caused long-lasting adverse effects by compacting the surface layers of the soil.

The effects can occur on soils of all textures and it is not generally realized that the subsoil can be compressed well below the point of application of the pressure, particularly in wet clay soils.

Lime and gypsum Normal liming operations benefit soil structure by flocculating the clay and causing it to aggregate. This can be regarded as a bonus to the effects of lime on nutrient availability, but on no account should overliming be practised in the hope of further improvements to soil structure.

Following the catastrophic flooding by sea water of large areas of East Anglia in 1953, there was an excess of sodium in the cation exchange complex of soils which dispersed and deflocculated the clays and adversely affected soil structure. Gypsum was used to replace the sodium by calcium and thereby restore the structure. It was preferred to lime because of its greater solubility, giving more rapid action, and also because, as a neutral salt, it would not give rise to overliming as lime applied at the necessary rates would have done.

There is, however, no point in using gypsum as a structure improver, except in these very unusual circumstances. Calcareous soils will already contain sufficient calcium to perform the job that gypsum would do and the amounts of lime applied as a routine to acid soils

supply more than sufficient calcium for maximum structural improvement without adding gypsum.

Synthetic soil conditioners A group of synthetic organic polymers was produced and marketed in the early 1950s. They stabilize soil structure by effects on the clay. Although they produced some remarkable and long-lasting benefits in field experiments they were prohibitively expensive. More recently polyvinyl alcohol and cellulose xanthate have shown promise in preventing soil capping. They are not, as yet, widely used.

Soil water problems

11

Excess water

Retention of excess water by soil reduces aeration, restricts root growth and may severely affect crop growth. When rainfall exceeds the requirements of the crop and those of the soil for storage of available water the excess water must be removed. In light-textured soils derived from sandstones, transported sandy parent materials, chalk and some limestones, the natural drainage can cope efficiently with excess water, which easily reaches water courses. These soils cover large areas of the British Isles. In England, for example, more than one-third of the total land area is occupied by well-drained soils, which do not need underdrainage.

The chances of excess water being present for all or part of the year are much greater on medium- and heavy-textured soils, on low-lying or flat sites, alongside watercourses, and in peaty soils which are by nature ill-drained. Problems will also be more common in areas with high rainfall and heavy-textured soils, which might be quite dry in East Anglia, can be excessively wet in north-west England.

If there is excess water at or within 40–50 cm of the soil surface during the cropping period, measures must be taken to remove it. It may simply be that old-established drainage systems have become blocked or outflows choked and some servicing is needed. There may be surface water caused by natural or man-induced pans which can be broken up by deep ploughing, subsoiling or even more drastic treatment with heavy equipment. The surface soil may have become puddled and capped by heavy rain through overcultivation and low soil organic matter.

Commonly, however, the problem is a high ground-water-table which must be lowered by the use of the tile, pipe or mole drains. There are one or two extreme conditions about which little can be done. These include some deep peats with very high water-holding capacity, heavy-textured soils in areas of high rainfall through which water moves too slowly, and potentially valuable heavy-textured alluvial soils in areas subject to periodic flooding which need major arterial drainage work covering a large riverside area.

The risk of excess water can be assessed by using basic soil maps or special-purpose maps (Fig. 4.2 and 4.4), but each case needs to be studied in the field. It is useful first to learn the drainage and cropping history of the site. Gradient and topography should be considered and, if possible, nearby soil exposures in quarries, pipeline trenches and building foundations should be examined.

On the site, if waterlogging of the surface horizons is frequent both leaves and roots of plants will show symptoms, some species being affected much more easily than others. The most susceptible crops are cereals which rapidly show symptoms of nitrogen deficiency. The leaves turn pale green, then yellow and the leaf tips and margins die, older leaves being affected most seriously. Plants are stunted and root growth is poor. If the plant is dug up and shaken, some rust-coloured soil may cling to the roots. Similar symptoms are shown by other crop species but only after longer periods of waterlogging.

Examination of the soil surface and the soil profile will be needed to identify the cause of the excess water.

Excess water in or on the surface soil can be seen, either as puddles or from the glistening appearance of the soil. There may be peat and sometimes a foetid smell when the soil is disturbed. More commonly, excess water will not lie on the surface but will rise and fall in particular horizons. If there is frequent excess of water in a horizon, root development in it will be poor, grey colours will predominate in the soil, the structure will be poor and there will be no earthworm activity.

The frequency of waterlogging can be judged from the nature and thickness of the gley horizons as indicated by the distribution of grey and brown colours. A general guide is given in Table 11.1.

The position of gley horizons in the profile and the relative greyness of different horizons are important in the diagnosis of the causes of waterlogging. A steady increase in greyness with depth indicates a Ground-Water Gley whereas a zone of maximum greyness in a horizon within 50 cm of the surface, underlain by a horizon with browner coloration indicates a Surface-Water Gley.

If there is doubt about diagnosis by colour, confirmation should be sought by monitoring the rise and fall of the water-table by dip-well or, preferably, piezometer readings. In some heavy-textured soils there is a delay of some days before the piezometer represents the true position of the water-table after rain, but if readings are taken regularly over a period of several months an accurate picture of the rise and fall of the water-table will be obtained.

Causes and treatment of waterlogging

Ground-Water Gleys

Ground-water problems occur in alluvial soils, especially in silty or clay-textured alluvium on the flat wide plains around slow-flowing rivers. Other sites include areas of lacustrine soils, glacial till and heavy-textured sedentary soil in which the structure deteriorates with depth. Ground water may also adversely affect crops on water-worked till or even on shallow fluvio-glacial soils underlain by heavy till.

There is a distinct rise and fall of the water-table as indicated by piezometer readings and this zone is mottled brown and grey (Fig. 3.3, page 35), greyness at depth

Table 11.1 Soil colour and frequency of waterlogging

Colour pattern	Frequency of waterlogging
Grey colours all-pervading	Permanent
Grey colours dominant, occasional rusty streaks or mottles occupying 10–20% of the horizon	Very frequent – probably all winter, with some root penetration during the summer
Approximately equal areas of grey and brown	Frequent during winter, infrequent during summer
Brown colours dominant, some grey patches	Occasional – probably insufficient to kill roots
Brown colours all-pervading	None

being progressively replaced by brownness towards the soil surface. In extreme cases the ground water is always at or near the surface. The mineral topsoil is grey and peat may develop.

The aim in this type of drainage problem is to lower the water-table as quickly as possible at critical times in order to create a greater rooting volume for rapidly growing crops. This is done by underdrainage. There are three types of drains used for underdrainage – tile drains, mole drains and, more recently, plastic pipe drains.

Tile and plastic pipe drains

Tile drains are cylindrical pipes of fireclay about 25 cm

long and varying in diameter from 10 to 15 cm according to the amount of water they will have to carry. They are laid in trenches, end-to-end and are led in patterns such as 'herringbone' into main drains of greater diameter and thence to ditches and natural waterways.

Plastic pipe drains are also hollow cylinders but the units are very much longer than those of tile drains, being up to 150 m in length. Water enters them through slits at the top of the pipe. There have been many arguments about the comparative effectiveness of tiles and plastics but they seem to be equally effective in most cases and decisions on their use depend mainly on cost. There are advantages in using plastic pipes in some peats where variable shrinkage caused by drying out can disrupt tile lines.

The use of permeable fill   If the soil which has been removed from the drainage trench is reasonably permeable it can be returned directly to cover the tiles. Infilling above the drains with heavy poorly structured soil can prevent water from reaching them. In this case some pervious material, such as gravel more than 6 mm diameter should be filled into the trench above the drains to provide a vertical channel connecting with the plough layer or with secondary mole drains (Fig. 11.1). This is expensive and can double the cost of a drainage scheme. None the less it can make all the difference between success and failure and permeable fill is now widely used.

Figure 11.1   Tile drainage with permeable fill

The use of filters Where drains are necessary in areas of light- or medium-textured soils with a high proportion of fine sand and silt and unstable structure, there may be problems of 'silting up'. Fine sand and silt are leached from the soil into the drains and may eventually block them. The problem can be tackled, rather expensively at present, by using fibreglass or some similar material to act as a filter by placing it over the drains when they are laid. Another method, used with some success in the Netherlands, is to bed and cover the drains with peat.

Unless one or other of these methods has been used desilting of drains will be necessary in these soil types every three to four years.

Depth and spacing of drains Drains must be laid sufficiently deep to avoid breakage by cultivation implements. The minimum safe depth is about 50 cm but more commonly drains are placed between 60 and 100 cm. Drains laid at the deeper part of this range are normally more effective provided there is an adequate outfall.

Traditional spacing of drains in many parts of the British Isles is very close, using intervals of 5–10 m. There is evidence that much wider spacing can be used, in some cases as wide as 40 m. The advisory services should be consulted about the appropriate drain spacing at a particular site and about secondary treatments such as subsoiling or moling.

Causes and treatment of waterlogging **147**

Mole drains

Mole drains consist of cylindrical tunnels, 7.5–10 cm diameter, created by drawing a special mole plough through heavy-textured soil. The plough has a sharp, vertical blade at the bottom of which is a 'mole', a bullet-shaped attachment which forms a channel, the walls of which consist of compressed soil (Fig. 11.2).

There are many soil types and conditions in which moling cannot be practised. The soil must be of heavy texture, stone free and must be stable against slaking by water. Moling is most satisfactory in uniform clay or clay loam soils and is quite useless in stony soils and in soils of light or medium texture where the tunnels would collapse rapidly.

Mole drains can be used in some heavy soils as complete drainage systems. The depth at which they are drawn depends upon the position in the profile of clay in which a stable mole channel can be established, but should certainly be more than 50 cm below the surface.

Mole drains are now more commonly used to provide temporary improvement to existing drainage systems. They are spaced much closer than tile or pipe drains, 2.5–3 m between channels being common.

Mole drains are cheap but short-lived, having a life expectancy of only 3–6 years, compared with plastic or tile drains which should, if well serviced, last for at least 50 years.

Figure 11.2 (left) Mole plough (N.I.A.E., Silsoe) and (right) subsoiler (S.I.A.E., Penicuik)

The use of pumping systems

The success of tile, plastic or mole drains depends upon having a sufficient slope to allow water to be carried away quickly after entry. Drainage of very flat areas such as the Humber estuarine alluvium, the East Anglian fens and smaller areas alongside some Scottish rivers depends upon the use of costly pumping systems to remove water from the ditches into which the drains lead. The amount of water which must be pumped out of the system to ensure that the drains work efficiently depends upon the difference between the annual rainfall and evapo-transpiration. In the East Anglian fens about 100–150 mm of water must be pumped out each year and this can be done economically in areas of intensive cash cropping. In areas of higher rainfall and lower evapo-transpiration more water must be pumped out. In the east of Scotland this can amount to 300–400 mm of water per annum, and pumping systems are unlikely to be economic.

Failed drainage systems

Before embarking on new drainage schemes it is essential to check whether there is a system in existence, if it has failed and, if so, why.

Few drainage systems can be expected to last indefinitely. Mole drains are drawn with the knowledge that they may collapse within a few years. More expensive tile or pipe systems are expected to be effective for much longer, up to 50 years, and they usually are. However, many drainage systems installed in the last century and early in this century are now becoming ineffective. There are several reasons for this:

- Cultivation pans may have formed, preventing access of water to the drains.

- Insufficient attention has been paid to outflows and ditches into which the drains discharge. This should be a regular procedure in poorly drained areas.

- Drains may have become blocked by fine sand or silt which gradually fills the tiles or pipes especially if gradients are small, the soil has a high fine sand and/or silt content and unstable structure and drain flow is not great.

- Drains may become blocked by iron ochre, an orange-yellow to red-brown rust-like deposit, usually colloidal in nature, which can accumulate fairly rapidly in drains from some types of soil, particularly acidic shallow peat with pyrite (FeS_2) or other iron compounds present in large quantities in the underlying mineral soil. In the case of newly reclaimed areas, drains may become blocked with ochre within three or four years, but the process is normally much slower.

Failure of newly installed drainage systems is not uncommon. Reasons for this include damage done by heavy equipment working in wet conditions during installation and the use of unsatisfactory materials for infilling, resulting in slow percolation to the drain. In both cases the problem could be anticipated and avoided.

Correction and prevention

Completely blocked, very old systems will need to be renewed. This will provide an opportunity to improve the design of the system by using modern techniques of spacing, materials and secondary treatment.

In all other cases the first step is to ensure that ditches and outflows are clear. If the main drains are then not flowing properly after heavy rain it will be necessary to check the drains for silting or ochre. The advisory services should be consulted on this.

Drain maintenance techniques have improved greatly in recent years. Both silt and ochre may now be removed by high-powered hoses, worked from the power take-off of a tractor, with a jet of water directed forwards to loosen the deposits and two jets directed backwards to sluice them into the outlets. The hose is introduced to the lower outlet of the drain and moves forward rapidly inside the drain. This method is more suited to the long plastic pipe drains than to tiles because, if it meets with an obstacle in a tile system, it is likely to cause an eruption and dislocation of the tiles.

Difficult drainage problems

High rainfall combined with heavy-textured soil

The natural drainage of heavy soils is very slow because of the high proportion of fine micropores. If such soils are compacted in any way drainage becomes even slower.

In these soils, the movement of water within the structural units is so slow that it does not help to drain away excess water. The water passes to the drains only through the cracks between structural units and hence the importance of maintaining these cracks. The presence of humus in the surface layers speeds up surface drainage, but inevitably the water draining away will meet with a water-table fairly near the surface.

In the drier parts of the British Isles artificial drainage can cope with the problem by causing structure cracks to open up and removing water through the shrinkage cracks but the wetter areas face a very difficult problem and drains placed at normal depths do not draw water. Crop yields are severely restricted if the water-table rises frequently to within 40–50 cm of the surface. For a clay loam soil in an area with 900 mm of annual rainfall this is likely to occur for the whole winter period and for one-third to one-half of the growing season. Such soils are, of necessity, usually kept under permanent pasture. Overstocking with cattle or sheep can cause serious poaching especially around food and water troughs.

There may be surface water and some peatiness in the surface soil. The surface soil, if not peaty will be grey in colour, with brown staining around roots. There may be remnants of ridge-and-furrow cultivations, the ridge being relatively dry but the furrow frequently having surface water.

The most that can be achieved in high rainfall areas is to remove surface water in some way. The soils do not respond to normal drainage systems even at the narrowest spacing, because of the very slow water movement.
Temporary alleviation can be achieved, provided that the stone content is low, by 'mini-moling'. This consists of drawing a mole about 5 cm in diameter through the soil at a depth of 20–40 cm and at intervals of 1–1.5 m. The moles will usually stay open for only three to four years and the process must then be repeated. Recently on drumlin soils in Eire gravel has been inserted in the mole to give it a longer life.

Liming, which helps to flocculate the clay and improve the structure, and the use of fertilizers to encourage vigorous growth and, hence, transpiration can help to dry out the top few centimetres but the economics and efficiency of high fertilizer use on such sites is always doubtful.

Avoidance of overstocking in wet periods is essential and many sites are suitable only for summer grazing.
Open ditch drains can be used at intervals of 20–50 m

with the aim of simply removing accumulated surface water, but the presence of these ditches severely restricts the type of agriculture that can be practised.

The drainage of peats

Once peat has formed its high water-holding capacity makes the problems of removal of excess water very difficult. This problem is not so serious on the fibrous and pseudo fibrous low moor peats of England and Wales as they contain fine colloidal fibres which respond to structure-forming influences such as liming. But in the peats of parts of Ireland and north-west Scotland the combination of high rainfall and high water retention of the resultant peat gives rise to a very difficult drainage problem.

The presence of peat is easily recognized in the field and its depth and type may be assessed by profile examination, although this would be a mammoth task for some peats, 10 m or more thick.

In general, the darker the colour of the peat and the more granular the appearance, the less difficult is the problem of drainage. Pale brown or yellow peats usually retain water very strongly.

Deep peat in dry areas such as East Anglia responds to deep ditch drainage provided that arterial drainage is satisfactory and the fields are not more than 100 m across.
The shrinkage of these peats resulting from oxidative losses

can be rapid and, unless anti-erosion measures are taken, wind erosion can occur [Chapter 13]. The land surface can sink by 1–2 m over a period of 50 years. When the peat has shrunk so far that it is only 1 m deep above the grey mineral soil, underdrainage becomes essential. Tiles used in such situations are likely to move because of continued uneven shrinkage and plastic pipes are preferable because of their great length. Care must be taken, if the peat is fibrous, to drain during a period of low rainfall, so that the trench is as dry and clean as possible. Putting down the pipes in slurried peat can result in the immediate blocking of the entry slits by the fibres and the drain will be inefficient from the time of installation. Fibreglass used as a filter over the drains can prevent this.

High moor and climatic peats   The drainage of deep high-moor and climatic peats is a major problem, both practically and economically.

The poorest sphagnum–cotton grass peats of Scotland and Ireland, for example, are not even responsive to ditch drainage because of their very high water retention. Shallower heather – purple moor grass (*Molinia caerulea*) peats which are more fibrous will respond to plastic pipe underdrainage but the costs are often prohibitive when considered in relation to improved output.

Some of these peats are used successfully for forestry. Others are harvested for fuel or for use in horticulture but many of them are extremely difficult to use.

Surface-Water Gleys

Many excess water problems occur because water cannot percolate through the soil to the ground-water-table but accumulates on the soil surface or in the upper horizons. The main causes of this are the breakdown of surface structure resulting in puddling and capping or the presence of horizons described as pans.

Pans

The term 'pan' applies generally to any soil horizon that is appreciably more compact, hardened or cemented and thereby less permeable than the horizons above or below it. There are several types of pan, all of which can restrict or effectively prevent the passage of water or roots. Naturally formed pans include iron pans, clay pans and hard pans. There are also cultivation pans, sometimes known as plough pans induced by the activities of man.

Pans present major problems in restricting crop growth. They commonly lie just below the cultivated layer and they prevent the downward movement of water giving Surface-Water Gleys. They restrict the effective depth of soil by restricting roots and, equally important, they restrict or prevent the upward movement of water to roots from below the pan, so that when crop growth begins the supply of available water can soon be exhausted.

A useful indication of the presence of pans in an area can be gained from windblow of adjacent forest trees; planted before the consequences were appreciated. These trees can grow to 5–10 m in height but roots will not have penetrated the pan. As a result, the trees may be blown down by strong winds, exposing extensive horizontal root systems running along the top of the pan.

The identification of pans

Iron pans   There are two types of iron pan, thin and thick. Thin iron pans are relics of podsolization in pre-cultivation eras. They are the product of very acid conditions and there is no chance of them forming in limed, cultivated soils. Many of these pans lying near the surface have been broken during reclamation. Others lie below normal cultivation depths and, if continuous can cause complete impedance of drainage.

Thin iron pans occur quite widely under natural conditions on light-textured soils in podsolized areas of the north and west of the British Isles. They are much less common in the south and east of England.

Examination of the soil profile will reveal a continuous, very hard, dark brown or black horizon, only 2 or 3 mm thick, at depths varying from 25 to 70 cm below the surface. There may be a diffuse orange-brown horizon below this and a greyish horizon above. Water will seep into the profile pit from immediately above the pan, at which level there may also be a mat of roots which spread horizontally because they cannot penetrate the pan.

The pan itself may be horizontal or may be very tortuous in shape (Fig. 11.3). If it is continuous it must be broken. If the pan is not continuous it, and the horizon below it, should be examined by probing with a knife or penetrometer. If there is appreciable resistance as compared with the horizon above the pan, remedial treatment may still be worth while.

Thick iron pans are much more intractable, but fortunately much less common than thin iron pans. They occur mostly in valley bottoms where the water-table rises and falls very frequently. They may be overlain by thin peat and are sometimes known as bog iron ore. The soil surface is seasonally or permanently wet and, at a point immediately below the cultivated soil or at greater depths the pan will be found – a strongly concreted brown or black horizon in which stones may be cemented (Fig. 11.4). These pans are at least 10 cm thick but may be 50 cm or more. A pick is needed to penetrate them.

Clay pans   In Brown Earths and Podsols there is movement of clay from the A to the B horizons. There can also be a physical leaching of silt particles through the mesh of sand grains in some soils with unstable structure. In many soils of light or medium texture the accumulation

Figure 11.3 A thin iron pan

of clay and silt in B horizons may only slightly retard the downward movement of water with no adverse effects on crops. It is in areas with soils of medium to heavy texture in which there has been appreciable accumulation of clay *below* the present cultivated layer that problems occur. Clay pans in arable areas are almost certainly inherited and are unlikely to develop where there is good liming practice.

Figure 11.4 A piece of a thick iron pan

During the digging of a profile pit there will be increased resistance to the passage of the spade when the top of the pan is reached and this will continue to the bottom of the pan. Clay pans vary in thickness, the usual range being 10–25 cm. In many cases there will be a remarkable change in the ease of digging at the bottom of the pan and this will be very useful in deciding the depth to which treatment is necessary.

As clay pans are not generally platy structured they will resist the entry of a sharp blade or penetrometer both horizontally and vertically to about the same extent. The pan will be heavier in texture than the horizons above and below and examination of structural units with a hand lens will usually show smooth clay skins on their surfaces.

Visual symptoms may be quite striking. There is gleying with increased greyness immediately above and in the pan with a sudden decrease in greyness (increased brownness) below the pan. Rooting will be very sparse in and below the pan. After heavy rain there will be seepage of water into the profile pit from above the pan.

Hard pans, indurated layers　These terms are used to describe horizons formed during soil development, which are hardened and sometimes cemented by silica or calcium carbonate. They also include layers compacted until the particles are very closely packed, perhaps by the pressure of very thick ice in northern Britain during the glacial period.

These pans will be revealed in the digging of the profile pit in the same way as clay pans but they are very much harder. They are very variable in thickness and in their position in the profile and are usually underlain by softer material, although they may easily be mistaken for rock.

Cultivation pans　These pans are entirely the result of human activity. They are widespread and becoming more so with the intensification of agriculture, the use of heavy machinery and the increasing tendency to cultivate and run over the soil when it is wet. They can form in soils of any texture but the most difficult problems are in sandy clay loams, clay loams and clays.

Cultivation pans cause serious loss of crop. Their formation is discussed, along with preventive and corrective measures, in Chapter 12.

Treatment of pans

The various types of pan described vary greatly in strength and the effort needed to break them varies accordingly. The implements used to break them are all designed to have a shattering effect.

Deep ploughing　This can be sufficient to break thin iron pans provided that they are within 40 cm of the surface. It can also cope with some thin cultivation pans, but care is needed to avoid ploughing when wet or a new pan might be created at the new plough depth. Deep ploughing is not recommended for breaking clay pans, because it is usually necessary to cultivate more deeply than 40 cm to be sure of getting through the pan. It will not cope with hard pans or thick iron pans.

Pan busting   There is an attachment called a 'pan buster', which is a small tine mounted immediately behind the plough and fixed to penetrate 10–12 cm deeper than the plough. It is designed to break up cultivation pans as they are formed but could be used to break up thin iron pans.

Subsoiling   This consists of pulling through the soil a strong shaft, sloping or vertical, to the bottom of which is fixed a chisel-like shoe sticking out in the direction of cultivation and set pointing slightly downwards at an angle of 20–25 ° to the horizontal (Fig. 11.2). There are many types of subsoiler but a typical shoe would be 10–15 cm wide and 40–50 cm long. The typical subsoiler can be set with the shoe operating at any depth down to about 60 cm and should be set some 10 cm below the bottom of the pan. The power requirement is great and this involves the use of heavy machinery with the risk of surface compaction if conditions are too wet.

The essential principle of subsoiling is to shatter the pan (Fig. 11.5), and to do this the moisture content of the

Figure 11.5 The shattering effect of subsoiling

subsoil must be low, below the plastic limit, in order to encourage the formation of structure cracks. If done successfully the surface soil will heave around the shaft of the subsoiler and there will be mounds along the lines of operation until the soil settles or is cultivated. The interval between runs varies but should be 1–2 m. When the subsoil is dry enough for subsoiling the topsoil may be even drier. This increases the draught requirement. Also the greater strength of the soil can prevent the heaving necessary in good subsoiling. In these circumstances it is wise to 'subsoil' twice, once to a depth of 25–35 cm and once to a depth greater than that of the pan. Alternatively, draught can be reduced by using two 30 cm tines on the same bar as the subsoiler but placed a few feet in front of it and a little to each side.

Periods during which the soil is dry enough for subsoiling are infrequent in the wetter north and west parts of the British Isles. It must be stressed that subsoiling when conditions are too wet is very inefficient and will, at best, create a smeared slit at the bottom of which will be a channel similar to a mole. In many soils the channels will collapse rapidly.

Subsoiling may be used to break thin iron pans, clay pans, and cultivation pans and will also cope with some hard pans. It may also be used as a general aid to permeability in drainage schemes on heavy-textured soils. Subsoiling for this purpose should be done at depths of 40–50 cm and intervals of 1.5–2 m. If this is done, the spacing of tile or plastic pipe drains may be increased. In all cases careful attention must be given to the depth and state of existing underdrainage and the subsoiler must be operated at depths where it will not damage drains. In contrast to the breaking of naturally formed iron-, clay- and hard-pans which is a once-for-all operation, subsoiling as an aid to permeability will have to be repeated periodically as the soil gradually settles. The time intervals between subsoiling operations will vary from soil to soil. The advisory services should be consulted before undertaking any subsoiling/drainage scheme.

The use of heavier implements Some thick iron pans, hard pans and indurated layers are so thick and tough that they cannot be broken by normal subsoiling. The only types of implement that would break them are the huge ripping machines used by engineers or some of the heavier 'prairie-busters' used in forestry operations. These are very expensive to operate and advice should be sought on the likely cost/benefits before using them. In many cases it will not be economic to attempt to break these pans.

Puddling, capping and poaching These processes resulting from surface structure breakdown commonly induced by cultivations (puddling, capping) or the trampling of stock (poaching) can cause water to accumulate on the soil surface. Their causes and treatment are discussed in Chapter 12.

Benefits of removing excess water

The benefits of rapid removal of excess water from soil are very great. Soil structure is improved and effective soil depth increased with consequent effects on aeration, root ramification and hence on the availability of nutrients to the plant.

There are other important effects:

Reduction of plasticity

The plasticity of heavy-textured soils is reduced as their water content decreases. The removal of excess water therefore helps to prevent cultivation pans, surface compaction and clod formation all of which result from cultivating the soil when it is plastic.

Increase in soil temperature

The higher the water content of the soil, the more heat is required to raise its temperature. (A given weight of water takes approximately five times the heat for a unit rise in temperature as does dry soil.) Thus under field conditions the water content determines, more than any other factor, the heat energy needed to raise the temperature of the soil. Soil containing 20 per cent of water will need 50 per cent more heat and soil containing 30 per cent of water almost 100 per cent more heat for a given rise in temperature than dry soil.

It follows that any excess water will slow down the rises in temperature necessary to initiate plant growth in the spring. It will also limit many temperature-dependent soil processes such as the conversion of organic nitrogen into available forms.

Frost heaving

Soil containing excess water in large pores can 'heave' at the surface on freezing due to the expansion which occurs. This is not a major disadvantage if the soil is bare but perennial crops, especially grasses, clovers and soft fruit, can be severely damaged by the fracture and exposure of roots to the atmosphere. Removal of excess water will reduce frost heaving.

Lack of water

There is no doubt that lack of water at critical periods of plant growth is a major restricting factor on crop yields in many parts of the British Isles.

The main causes are:

- Insufficient rainfall during the growing season.
- Erratic distribution of rainfall from day to day or month to month.
- Low available water capacity of the soil.
- Restricted water movement in the soil, brought about by pans or heavy texture.

Lack of water is most acute in sands, loamy sands and coarse sandy loams, the available water capacity of which is low, but can also occur in soils of medium and heavy texture especially if the structure is poor.

If the water-table is effectively out of the root range of crops (2 m +) as in many areas of light soil, the combination of rapid percolation, low available water capacity and unsatisfactory rainfall distribution leads to frequent soil-water deficits, particularly in south-easterly parts of England and eastern Scotland.

Available water capacities are shown in percentage terms for various texture types in Fig. 7.2 (page 91). When assessing the likelihood of drought (conditions in which lack of water seriously reduces the yield of crops) these figures are expressed as millimetres of water for a given depth of soil, usually 90 cm. The available water capacities of some texture groups, expressed in these terms are given in Table 11.2. The available water capacity of sands, loamy sands and sandy loams is in the range of

Table 11.2 The available water-capacities of some soil texture types

Texture type	Available water capacity (mm water/90 cm soil depth)
Sand	60–80
Loamy sand	80–100
Sandy loam	100–140
Loam	140–180
Silty loam	140–200
Clay	130–160

60–120 mm per 90 cm of soil depth. The needs of crops for transpiration are some 20–30 mm per week and it is easy to see that any water reserve will be rapidly used up unless it is replaced by rainfall or irrigation even if roots have access to the whole 90 cm of soil depth.

All areas of the British Isles have an excess of precipitation over evapo-transpiration during the period October–March, so that soils will be at field capacity when crop growth begins. They will then begin to dry out as transpiration proceeds.

The term 'soil-water deficit' is used to represent the difference between the crop requirement for optimum growth and the summer rainfall. This can be calculated from the meteorological data for the current year and tables of 'potential transpiration' available from the Ministry of Agriculture, Fisheries and Food (MAFF). The

Table 11.3 Number of 'drought' years in 10 years

Area	Drought years/10
South-east Scotland	1–3
North-east England	1–3
East-midland England	3–5
West-midland England	2–3
South-east England (inland)	5–6
South-east England (coast)	7–8

only source of water to offset the deficit is that stored in the soil as measured by its available water capacity.

A simple way of expressing 'drought' is the arbitrary use of a soil-water deficit greater than 150 mm during the period April–September. On this basis Table 11.3 shows the number of years out of 10 in which drought occurs in various parts of England and Scotland. Even on this basis, light-textured soils in east and south-east England suffer from drought in 8 years out of 10. The actual position is worse than that because no account is taken of drainage losses following heavy rain, the uneven distribution of rainfall or of the variations in water requirements of crops at various stages of growth. For example, the most critical period of water requirement for cereals is immediately after ear emergence. In the drier parts of the country this will occur from mid-May onwards. Potatoes, on the other hand, need most water during the extraordinary period of growth when the tubers are swelling (June to August depending on variety and area). Figure. 11.6 shows the incidence of drought for the early and late growing season, April–June and July–September, assuming that a soil moisture deficit of 75 mm for either period represents drought. Drought during April–June is more frequent than during July–September. In fact the whole of south-east England experiences early summer drought five years out of ten, at a time of year critical for the water requirements of cereals and the grass crop.

The actual lack of water for transpiration is the main effect of drought but another important aspect is the reduction in availability of nutrients to the plant. Most plant nutrients are taken up from soil water. Also many of the processes by which nutrients become available to the plant require water. Mineralization of organic matter is a good example. The feeding rootlets of plants die unless they are bathed in water and usually the first part of the soil to dry out during drought is the cultivated layer where both roots and available nutrients are concentrated. In these circumstances the plant may be able to draw on water and nutrients only by sending roots further down in the soil where nutrient concentrations are low.

In extreme cases, lack of water is obvious from the wilting of plants and the dry appearance of the soil. The growth of crops is adversely affected long before this

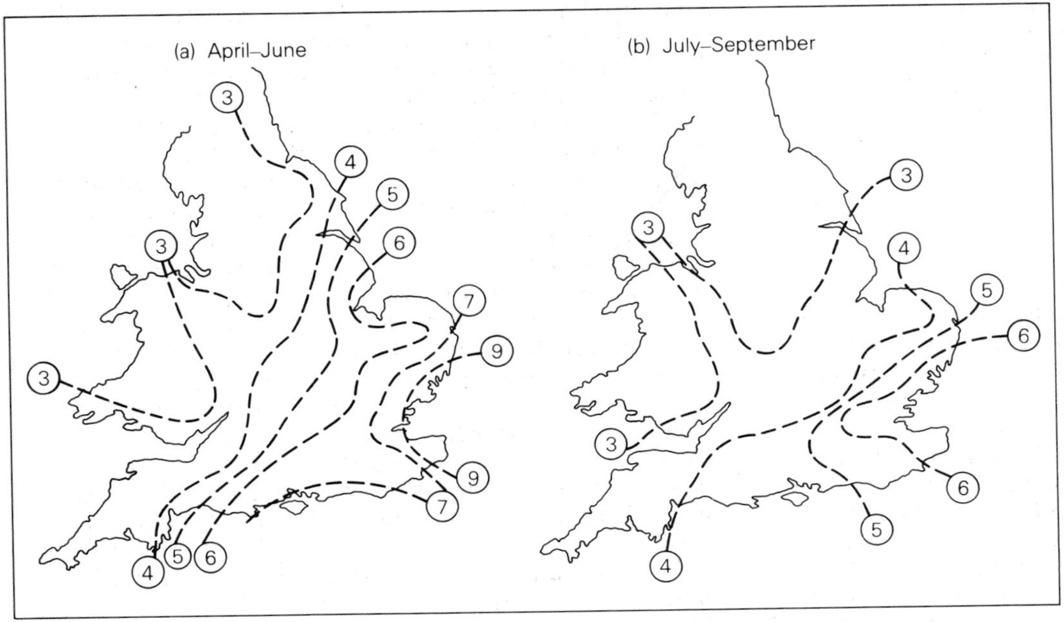
(a) April–June
3
3
3
4
5
6
6
7
9
9
7
4
5
6
(b) July–September
3
3
3
4
5
6
6
5
4

Figure 11.6 The incidence of drought in England and Wales. Numbers show years out of ten in which drought occurs according to definition on p. 160

because of the restricted water supply as affected by root ramification and speed of water movement through the soil. Also plants close their stomata as drought approaches, cutting down the amount of water transpired, thereby reducing the efficiency of the plant.

Irrigation

The only effective way of supplying water to soils with a deficit is by irrigation, which is costly in terms of labour, equipment and water supply. Furthermore water is usually scarce in areas of high irrigation requirement and the farmer must decide whether to irrigate or not on the basis of access to water and costs set against the returns from extra crop yields.

Risk of drought in a particular soil can be assessed by using meteorological data for the area and the available water capacity of the soil as determined by its texture, and the need for irrigation estimated accordingly. It is impossible in a book of this type to give details of the calculations but the following factors are taken into account when assessing irrigation needs:

- Available water capacity of the soil.
- Potential evapo-transpiration on a weekly basis.
- Crop to be grown.
- Percentage soil cover by plants in early stages of growth.
- Time at which soil cover becomes complete.

Using these factors, rule-of-thumb guides to irrigation requirements can be produced depending upon the responsiveness of crops.

For example it is not generally recommended to irrigate potatoes until the tubers begin to swell, or sugar beet until the leaves meet between the rows. Brassica crops for human consumption, on the other hand, should be irrigated from the time of planting and grass can benefit from irrigation throughout the growing season.

Irrigation plans are usually based on estimates, derived from field experience, of what water deficit a particular crop will stand, without adverse effects on growth, in soils of various available water capacities.

In practice many crops will go without irrigation. Even in dry intensive cash-cropping areas where growers are almost obliged to invest in irrigation equipment, they are faced with deciding which crops to irrigate. For instance, cereal crops are rarely irrigated although yield responses could be spectacular.

Despite the known drought risks for very large areas of the country, shown in Fig. 11.6, irrigation is practised only in limited areas, and quite rarely in those parts where drought risk occurs in 1, 2, 3 or 4 years out of 10. There are several reasons for this. One is the cost of buying, storing and using equipment which depreciates during non-drought years. Another is the accessibility and cost of irrigation water and a third is the labour requirement for the operation of the equipment.

Even in areas where drought is virtually a yearly risk a large proportion of crops goes unirrigated. It is very unfortunate that this leads to inefficient crop responses to lime, fertilizers, weed control and careful cultivations.

It is a sad reflection on agricultural planning in the United Kingdom that, despite the self-sufficiency needs experienced during two world wars, we have no water-grid system that could cope with irrigation requirements. Urban-dominated electricity, gas and restricted water grids for domestic use have met with little resistance and more recently oil and natural gas pipelines have proliferated but the need for an irrigation-water grid is ignored. In the United States of America irrigation water is carried through mountains, run for hundreds of miles and applied at extraordinary rates sometimes exceeding 300 cm per year. In the United Kingdom about one-third of the land area has a gross excess of rainfall and about one-third has a gross deficiency. In national terms, the costs of establishing storage in high rainfall areas and pipelines from, for example, north-west Scotland to East Anglia would be small and would be even more effective than the lime subsidy which boosted crop yields enormously between 1940 and 1960. It is odd that the farming community, so effective as a lobby in many ways, mutely accepts the lack of water which is the greatest restriction to crop production in this country.

Reducing irrigation needs

Irrigation needs can be reduced to some extent by increasing the available water capacity of soil. As shown in Fig. 7.3 (page 92) this can be done by building up the organic matter content and improving the soil structure by methods described in Chapter 10. In areas where drought occurs frequently, building up organic matter is, however, difficult and slow because of oxidative losses from dry soils.

Problems arising from cultivation and stocking

12

Cultivation problems and problems caused by overstocking can occur on soils of widely differing type and texture. Most of them can be prevented by careful management of the soil but some, especially in wetter areas, are very difficult to overcome.

Loose or 'puffy' seedbeds

Cultivation of light-textured soils is, on the face of it, so easy that the seedbed problems associated with these soils are often neglected by farmers who have greater cultivation problems on their heavier soils. Large areas of light land can be ploughed, cultivated and sown within a day or two and ploughing is often delayed until the last minute. The soil has no time to settle and this leads to puffy seedbeds that are too loose and insufficiently compacted before sowing. One consequence of this is deep drilling of cereals. Some seedlings have to struggle 6–8 cm to the surface before emerging.

Frequent runs with machinery or rollers over the ground in attempts to correct the puffiness before or after sowing are seldom successful and can cause pan formation.

Puffy seedbeds can be avoided by ploughing early enough to allow the soil to settle naturally. This is easily done for spring sown crops by ploughing in the autumn but is not possible for autumn sown crops such as winter barley and winter wheat. The alternative is to use minimal tillage or direct drilling systems in which the soil surface is very lightly cultivated or the seeds are sown directly into the stubble without any cultivation at all. Light soils are very good media for direct drilling. There is a considerable saving in energy and organic matter is conserved but attention must be paid to control of perennial weeds such as couch grass (*Agropyron repens*) by the use of herbicides.

Puddling and capping

Low organic matter soils with high proportions of fine sand or silt can be puddled by heavy rain. Puddling, followed by capping, is most likely to occur when the soil has been cultivated to a fine tilth and when the surface is free from crops and weeds. The most critical time is immediately after sowing and caps formed after seedling emergence are not so important. Puddling is caused by the breakdown of the surface structure by the onslaught of raindrops on bare soil. The splash from the raindrops carries fine particles which are deposited in a thin layer, only a few millimetres thick, on the soil surface. Puddles then build up and the fine particles sediment out to form a mini-profile of parallel horizontal layers with the finest particles at the top. The puddles cannot drain away as the layers are

strongly bonded and become more so as the soil dries out by evaporation. This is known as capping or crusting. Capping is not a serious problem so long as the cap remains moist. It is the great increase in strength on drying that causes problems. If, however, the cap remains wet for some weeks, conditions under it become anaerobic and damage to seedlings can result.

Capping is not important in clay soils but can be a problem on light soils with a high fine sand content and is most serious in silty loams and silty clay loams. Caps formed on fine sandy loams or fine loamy sands, although not so strong as those on silty soils, can interfere with the emergence of small-seeded crops such as carrot, onion and lettuce. Caps formed on silt can impede the emergence of cereals and even potatoes or bulb crops. It is very unwise to use precision drilling 'to a stand' of sugar beet or other small-seeded crops where there is any risk of capping.

Capping, although serious in its effects, is a temporary condition. The caps are easily destroyed by light cultivation but can form again at the next heavy rain. Some alleviation of the problem can be achieved by cultivating to leave the soil surface as rough as is consistent with sowing the crop. It is also helpful to leave stubble and crop residues on the soil surface, using 'stubble mulch' cultivations which incorporate them very lightly.

Maintenance or improvement of organic matter levels is critical in the prevention of capping. Many of the rich silty soils susceptible to capping are, however, intensively cropped and methods of increasing organic matter which reduce the time available for cash cropping are unacceptable to the farmer.

For example they will be unwilling to resort to grass leys. Also, because livestock are not numerous in these regions, bulky manures are not available unless transported unacceptable distances.

Green manure crops, grown at any opportunity, are the most satisfactory method of getting organic matter into these soils, and green crops serve the further purpose of maintaining cover when cash crops are not being grown. Great care must be taken on silty soils to avoid the creation of anaerobic layers when incorporating green crops. The crop should not be ploughed in but should be vigorously chopped or bruised and incorporated lightly into the surface soil.

Surface compaction

Surface compaction can occur as a result of almost any machinery operation and will be worst if any of them is done when the soil is wet. Tractor wheeling, loaded trailers and harvesters give rise to most of the problems. For example, tractor wheel tracks during traditional seedbed preparation can cover as much as 90 per cent of the soil surface. Late, 'wet' harvests of crops, especially potatoes and sugar beet also cause serious surface compaction.

The result of surface compaction is to reduce total pore space and to increase the proportion of finer pores. It can occur on soils of all textures but most susceptible are soils with high proportions of fine sand and silt which pack together readily and heavy-textured soils cultivated when too wet.

Compaction in light soils should not be neglected. For example, attempts to correct puffy seedbeds by repeated runs of machinery can, especially if the soil is wet below the surface, cause considerable compaction within and immediately below the plough layer. This condition, in which the surface soil is dry down to 10 cm or so and wet below, occurs quite frequently in light soils. Compacted layers in light soils are insidious as they are not easy to identify by probing the profile with a knife blade and the soil in these layers crumbles easily when crushed in the hand but they do stop root penetration and this is critical in soils of such low available water capacity.

Their presence is best identified by observing root patterns. In cereals, roots become 'stubby' and branched. Tap-rooted crops such as sugar beet become deformed and the growing tip may be lost, resulting in fangy roots and lateral roots may be run for 15–25 cm horizontally. Other crops also develop horizontal rooting patterns. Most seriously affected of the light soils by surface compaction are loamy sands with a high proportion of fine sand.

Of the heavier soils, silty loams and silty clay loams are most susceptible, but all heavy soils can be affected. Because of their high water-holding capacities the silty loams easily become anaerobic if compacted. The problem is worst when fresh organic matter, such as sugar beet tops, has been recently incorporated by complete inversion of a very wet furrow slice. Toxic gases (hydrogen sulphide, ethylene) are produced and there is evidence that denitrification is also vigorous. As a result crops can suffer from nitrogen deficiency and, as in lighter soils, fangy roots may be found in root crops.

Another effect of surface compaction on heavy soils is the formation, during subsequent cultivation, of large clods with a large proportion of fine micropores. The clods are sticky and cohesive when wet and tough, hard and difficult to break down when dry and this can lead to further damage because of the many passes that are needed to get a surface tilth when the soil below is still wet. The clods vary considerably in strength, those formed in compacted low organic matter clay soils with non-expanding clays (kaolinite) being most difficult to break. A moderate proportion of small clods of 5–10 mm 'diameter' is not

detrimental but tilths with a high proportion of clods greater than 50 mm make a very poor medium for germination and seedling emergence.

Surface compaction may be reduced in soils of all textures by reducing the number of passes made by tractors and other machines, by using the lightest machines possible, by spreading the load on the soil by the use of suitable wheels and tyres and by keeping off the land when it is soggy and wet. It is critically important not to be deceived by apparent dryness in the top 10–15 cm (Fig. 12.3). The damage is done in the wetter layers below. Excessive cultivations should be avoided both to reduce compaction and to avoid oxidative loss of organic matter. For example the number of seedbed passes on light- and medium-textured soils should be minimal, leaving coarser and less dense tilths. In silty soils, tractor wheel slip caused by the low-friction silky texture can be a serious cause of compaction. Wheel slip can be reduced by not allowing the implement draught to approach the maximum draw-bar pull and by avoiding tracking when the soil is wet.

As with so many soil problems the maintenance or improvement of organic matter levels is critical. Well-humified organic matter helps greatly to cushion the soil against compaction. This is clearly shown in Fig. 12.1 in which air and water-filled pore space are shown for a silty loam compacted by wheel tracks.

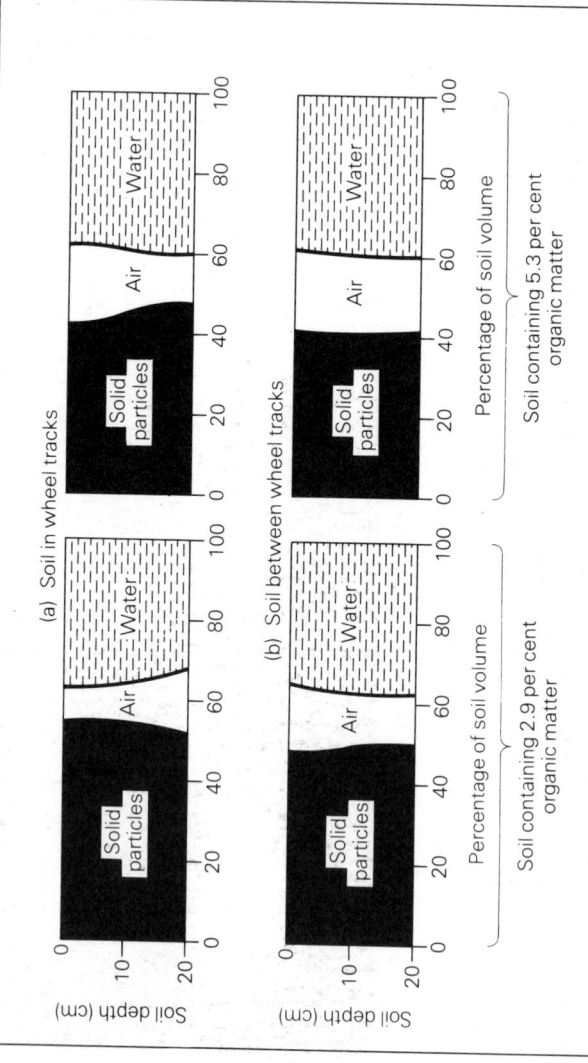

Figure 12.1 Effect of humified organic matter on air and water-filled pore space. (Derived from data in 'Field behaviour of medium textured and 'silty' soils.' D.B. Davies in 'Soil Physical Conditions and Crop Production.' Technical Bulletin **29**, M.A.F.F. 52–75 1975)

Cultivation pans

The terms 'cultivation pan' or 'plough pan' refer to any compacted and/or smeared layer lying immediately beneath the cultivated layer. The pans can be quite thick – as much as 20 cm – or may be only a centimetre thick (Fig. 12.2). As with surface-compacted layers, cultivation pans can occur in soils of any texture. They are most prevalent in medium- and heavy-textured soils in which ploughing or rotavating has been done to the same depth for many years.

The pans can completely prevent root penetration and a mat of roots spreads horizontally along the surface of the pan. This, combined with the slow upward movement of water through the pan, can cause drought conditions with daytime wilting of crops such as sugar beet. Growth of all crops can be retarded and nitrogen deficiency symptoms are sometimes shown. The yield of cereals and root crops can be seriously reduced and crops such as peas which are very sensitive to both excess and insufficient water cannot be grown at all.

Cultivation pans result from cultivating the soil when it is too wet. In medium-textured soils they are caused mainly by compaction beneath the plough sole but in heavy soils there is the added disadvantage of smearing or polishing the soil immediately below the plough which seals off the natural drainage channels.

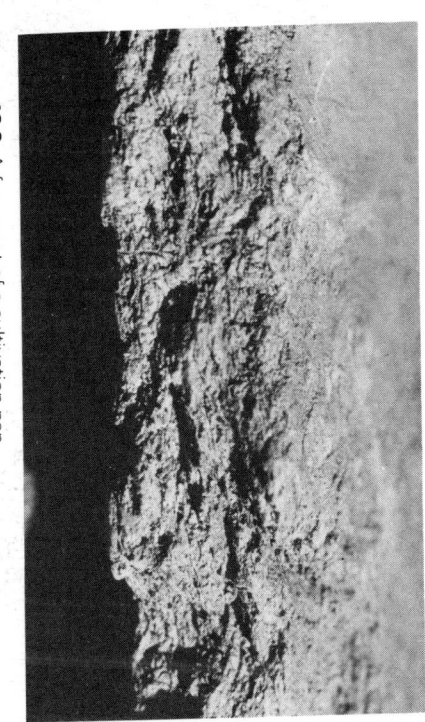

Figure 12.2 A fragment of a cultivation pan

The critical factor in cultivation-pan formation in heavy soils is the plastic limit – the percentage of water that the soil must contain to become plastic. Great care must be taken before ploughing to make sure that the soil at the bottom of the plough layer is below the plastic limit. Very sharp changes in moisture content with depth can occur when the soil is drying during the spring and the top few centimetres will be dry when the soil below is wet and plastic. This is shown in Fig. 12.3 for a clay loam soil. In winter the whole soil is plastic and, although the top 5 cm

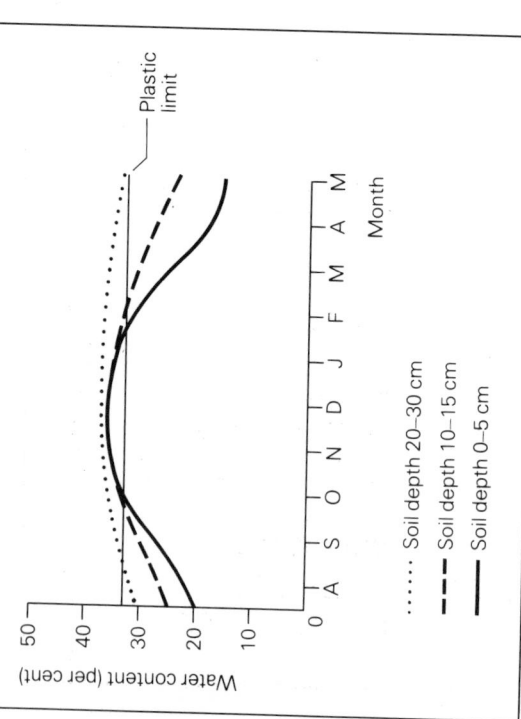

Figure 12.3 Water content and plastic limit of a clay soil over a period of several months

the country the moisture content of heavy soils at the foot of the plough layer is such that they are always plastic during the spring cultivation period. Arable cropping in such soils must be undertaken with a full knowledge of the risks and consequences of pan formation. Fundamentally they should be kept under pasture.

Plough pans may be identified in the soil profile by the methods used for other pans (Chapter 11). The main difference is that plough pans are usually strongly platy for at least a few centimetres below the plough layer. Because of this there will be more resistance to the passage of a sharp blade or penetrometer vertically than horizontally.

Cultivation pans do not have the strength of hard pans or some clay pans and can be broken by subsoiling or in some cases simply by deep ploughing when the soil is dry enough to be non-plastic.

A 'pan-buster', consisting of small subsoiling tines set immediately behind the plough, has been used in attempts to break up cultivation pans at the same time as they are formed. The tines are set to a depth of 10–12 cm below the plough. These implements add considerably to the power requirement. Also they can succeed only if the soil is dry enough to achieve the necessary shattering effect.

After the pan has been broken the drainage system should be inspected to see if it can be improved. This would help to prevent new pans from being formed.

becomes non-plastic and could be cultivated in March, the lower part of the plough layer and the subsoil do not lose plasticity until after crop growth has begun in May. Thus ploughing this soil at any time, winter or spring, is certain to cause pan formation. Unfortunately, in the wetter parts of

'Timeliness' of cultivation

Wherever arable crops are grown, it is essential to avoid trafficking the soil when it is too wet. Problems of clod formation, surface compaction and panning make this particularly important when dealing with heavy soils.

This can put the farmer in an almost impossible position in some years. Figure 12.4 gives an example of the 'number of working days' available for cultivations in an area of moderate rainfall during a dry spring and a wet spring. The average number of working days for the heavy soil during the critical months February–April is 40 and, for the light soil 60. The number can fall to only 20 during a wet spring. Obviously in high rainfall areas the position would be even worse.

The role of humified organic matter in increasing the number of working days by increasing the plastic limit of a heavy soil is shown starkly in Fig. 12.5, once again demonstrating the need for organic matter maintenance. Fortunately it is easier to maintain or increase organic matter levels in heavy soils than in medium- or light-textured soils.

There are a few soils which are exceptions to the general need for organic matter. In the drier parts of southern England, for example, some calcareous clays such as the Lower Lias can be managed under mainly arable rotations, without resorting to special measures for organic matter

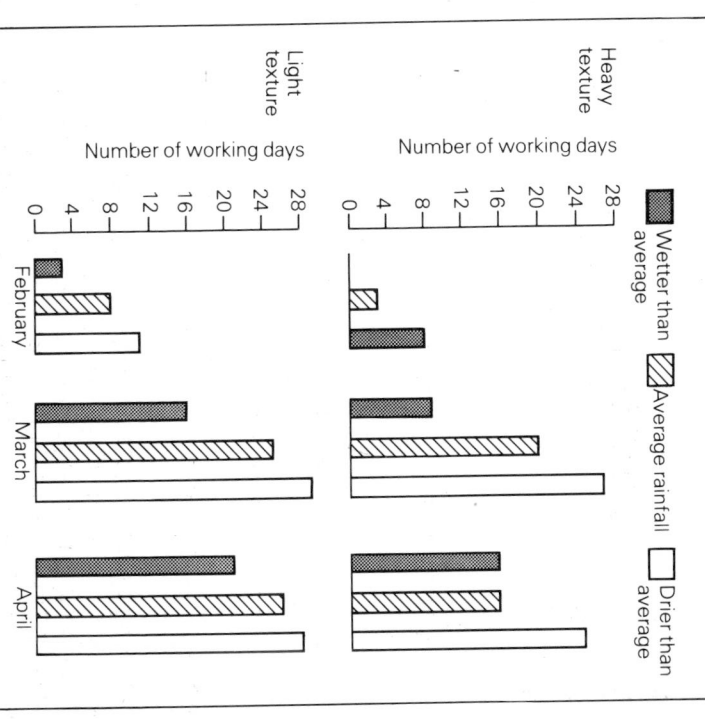

Figure 12.4 Number of working days available for cultivations

On all other heavy soils organic matter conservation is essential.

Tracking

All tracking, that is the passage of any implement over the soil, causes some compaction. The amount of tracking for arable crops has increased very considerably in recent years with the advent of insecticide and fungicide sprays. It is not inconceivable for a field to be tracked some 12 times, not counting ploughing, during the growing season. Recently there has even been a move to spray against couch grass by running through a standing, ripe cereal crop.

Much of this tracking is done in a haphazard manner, especially in harvest operations. The problem is made worse by the large variety of wheel base sizes, spray boom sizes, width of coverage of fertilizer distributors and numerous other variations.

Direct damage to crops by tracking is perhaps most serious in close grown cereals, but considerable damage to soil structure can occur in fields carrying row crops.

Some growers of high-value crops have adopted 'bed' or 'tramline' systems in which all traffic follows parallel tracks on which no crop is sown. There is an excellent case for extending these systems to all crops. This would require

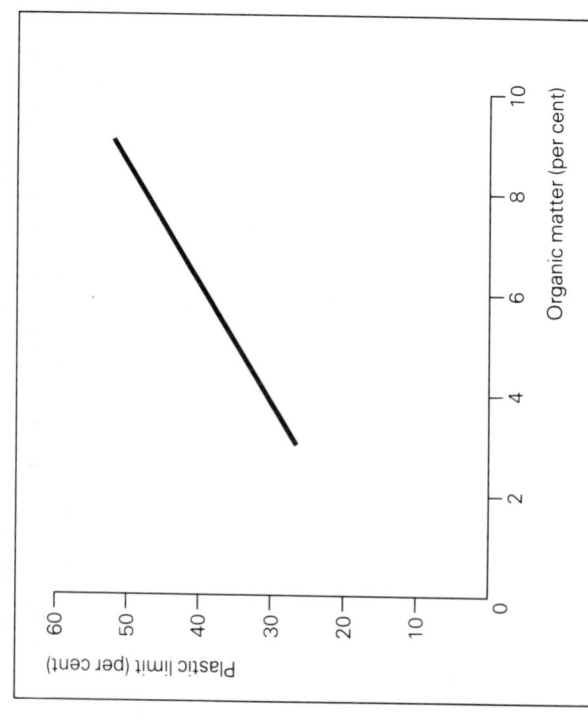

Figure 12.5 Organic matter content and plastic limit of a clay soil

maintenance, because of their remarkable structure – recovery properties. Even in these soils care is needed to retain the 'summer tilth' by using a 'no-ploughing' system or a minimum cultivation technique with tined implements.

major efforts and discipline on the part of machinery manufacturers and farmers but would give very large rewards in reducing compaction and crop damage.

Poaching

The term 'poaching' refers to the damage done to surface soil structure by the hooves of animals. It occurs mainly in wet conditions and is by no means confined to heavy soils. It can, for example, be quite serious on gleyed light-textured soils if overstocked. The major problem, however, is on heavy soils under grass, overstocked by cattle and to a lesser extent sheep, when the soil water content is above the plastic limit.

Poaching can damage both the herbage and the soil so much by puddling and compaction in the depressed footprints of the stock that reseeding becomes essential. It can produce pans, 10–15 cm below the surface, similar in nature to cultivation pans but not so continuous. Once formed these pans hold up water and create conditions still more likely to lead to poaching. In some high rainfall areas, with heavy soils, the problem is so serious that the land can be used for grazing only in the summer.

The obvious way to avoid poaching is to withhold livestock from wet soil. Zero grazing systems, in which the animals are kept off the fields and the grass is taken to

them either direct or as silage, are helpful in some cases. Attention to both underdrainage and surface drainage is essential. Subsoiling will be effective provided that a period can be found when the soil is dry enough.

Erosion problems

13

Soil erosion is the removal of some or all of the soil in an area, by wind or water. It is almost always the result of adopting management systems that are unsuitable to the soil and climate of the area. This has been tragically illustrated by the dust bowls created in the United States of America, in Russia and in several other parts of the world following the ploughing out of large areas of natural grassland. Overgrazing by wild or domesticated animals can also initiate erosion. Striking examples of this can be seen wherever rabbits colonize hill slopes.

On a world scale the problem is massive – nearly two-thirds of the whole area of the USA has been affected to some degree.

Soils in cool humid regions are less susceptible to erosion than those in drier, warmer parts of the world. Erosion in the British Isles is seldom catastrophic, but it is sufficiently serious in many areas to require preventive and corrective measures.

Wind erosion

Wind erosion is a problem with all sandy soil types. Some peats become erodable when they have dried out as a result of drainage, cultivations and intensive cropping. Problems can also occur with silts.

The size of unaggregated particles most likely to 'blow'

is 0.1–0.2 mm, in the fine sand range. Both bigger (up to 0.5 mm) and smaller (down to 0.05 mm) can 'saltate'. Saltation is the process by which much wind erosion occurs and is, effectively, the repeated bouncing of particles along the soil surface and the initiation of bouncing of other particles. Most of the particles in a 'blow' by saltation are no more than 20–30 cm above the soil surface although anyone who has walked through an area eroding in strong wind will know the discomfort to the face 150–200 cm above the ground. Much of the material at this and greater heights consists of silt and clay particles, which are borne directly in the air and do not saltate.

Wind erosion, once started, is likely to go on. Wind-blown soil which has been pegged down by vegetation and is then ploughed will probably blow again. There are high risks on many cultivated Podsols on fluvio-glacial sands or sedentary soils derived from finely grained sandstones, especially in dry springs before crop cover is established. Peats which have dried out following drainage and have been intensively cultivated to a fine tilth are also very susceptible because of the low density of the particles, which begin to shift in relatively light winds.

Wind erosion is exacerbated by lack of shelter, excessive cultivations, intensive use of herbicides, dryness and, in mineral soils, weak structure caused by low clay content and lack of organic matter.

The most likely period for wind erosion is immediately before or after sowing at the time when row crops are newly emerging in the sping. The consequences can be serious. They include loss of valuable topsoil and recently applied fertilizer, loss of seeds and germinating seedlings, injury to established crops during the blow and the blocking of ditches in which the soil inevitably collects.

Preventive and corrective measures

Remedies for wind erosion are based mainly on reducing wind speed and on building up soil structure. Measures that can be taken will vary according to site and the value of the crops to be grown.

The cheapest and simplest methods include leaving stubble and crop residues on the soil surface and using 'stubble mulch' cultivations by, for example, light reciprocating harrows. Both this and traditional cultivations should aim to leave as rough a surface as possible. Maintenance of continuous hedgerows in good condition will help to reduce wind speeds.

In many cases more expensive and time-consuming measures are required. The growing practice, during the last 30 years, of removing shelter belts and hedges, in the interest of creating larger, more easily handled fields has contributed beyond all measure to wind erosion risk. This trend must be reversed.

There is an urgent need to grow new shelter belts

designed to allow about one-third of the wind to pass through. A general rule is that shelter belts give protection for a distance 20 times their height. Completely impenetrable shelter belts can cause turbulence on the leeside and may make erosion worse.

Until the living shelter belts are established temporary artificial windbreaks, such as hessian or woven plastic can be used.

Other useful measures include the use of bulky organic materials for surface mulching and the use of rapid growing 'nurse crops' between the rows of valuable crops. The nurse crop should be one like mustard, which can be killed once the main crop is established. This serves the fourfold purpose of pegging down the soil by root ramification, green manuring, mulching, and reducing wind speeds at ground level.

In extreme cases resort can be made to 'strip cropping', so widely practised in erosion-prone countries. Easily damaged crops are grown in strips 10–20 m wide, interspersed with similar strips of cereals which are more resistant to erosion. This gives rise to difficult farm management problems.

Another extreme measure is to change the texture of erosion-prone soils by claying or marling. At least 400 t/ha of clay or marl must be added and intermixed with the surface soil. This would be feasible only if clay or marl is

available nearby and economic only if high-value crops are grown.

There is no doubt that the long-term strategy for the prevention of wind erosion should be based on a combination of permanent shelter and improving soil structure by means of increased humus content.

Water erosion

Water erosion is caused by rapidly moving water. The start of water erosion depends upon the severity of impact of raindrops, the rate of precipitation which encourages surface run-off and the slope of the land. It is very heavy storms which cause most erosion because the raindrops are larger and are moving more quickly than those in light rain.

If the erosion is slight it usually results in clay and silt particles becoming suspended in surface water and moving downslope in a mass. This is called sheet erosion and, if mild, usually leaves behind it a cap of fine particles.

If the erosion is more severe, water running off a slope will concentrate into tiny streams which may deepen and widen to give rill or even the catastrophic gully erosion, very rare in the lowlands of the British Isles. Simple examples of rill erosion can be seen on soils treated regularly with herbicides to eliminate weeds. This, while otherwise desirable, increases the direct impact of raindrops and removes the binding weed roots. In these circumstances small rills can occur in some weakly structured silts and clays when the slope is only 2–4°.

Water erosion is not a problem in sands, loamy sands and coarse sandy loams because of rapid infiltration rates. There can, however, be problems on light soils with high fine sand contents which, because of heavy storms, go beyond the pudding and capping stage to give sheet or rill erosion.

The most serious problems are found on medium-textured soils and on poorly structured clay soils. Most clay soils have some resistance to water erosion because of their cohesive properties but low-organic medium-textured soils rich in silt or fine sand are vulnerable because of the ease with which these particles move in flowing water.

Water erosion can occur in any season, depending, as it does on heavy, prolonged rain.

Prevention

Water erosion is seldom serious enough in the British Isles to warrant the major defensive measures that are necessary in parts of the world where torrential rain is common. Where there is a risk, sensible husbandry measures will minimize it. As in the case of wind erosion, stubble mulching is helpful. Ploughing and the drilling of crops like potatoes and ridge-grown swedes should be done along, or preferably just off, the contours instead of up and down the slope which creates ideal starting conditions for water erosion. Figure 13.1 shows the result of erosion, in a field with sandy clay loam surface texture, where potatoes were planted up and down a 7° slope except for a few drills at the foot which were subsequently broken down by the force of the water.

Attention to drainage, subsoiling and the avoidance of breakdown of surface structure by machinery will all help the infiltration of water and thus help to combat water erosion.

Figure 13.1 An example of water erosion (Photograph: G. Finnie)

Chemical problems

Part three

Diagnosis of chemical problems

14

The plant, in order to grow efficiently, needs to be able to take up essential nutrients virtually 'on demand'. If it cannot do so, or if it takes up an excess of various nutrients, it will not thrive.

Chemical problems in the soil which may cause poor plant growth may be diagnosed by using inferences from the way in which the soil has been formed, the examination of both crops and soils in the field and the analysis of plants and soils in the laboratory. Also the quantity of one element in the soil may affect the uptake by the plant of another. The most effective diagnosis requires, therefore, a different combination of methods for each nutrient element.

Successful diagnosis depends, therefore, as it does with physical problems, on bringing together facts from various sources and interpreting them as a whole. Figure 14.1 gives a guide to the soil and plant symptoms of deficiency of essential elements.

Soil formation and type

The type of soil and the way in which it has been formed can give valuable clues to the risk of deficiency or excess of various nutrients. This has been described for both sedentary and transported soils in Chapter 2. For example, the history of continuous leaching in a Podsol derived from

Figure 14.1 Plant and soil symptoms of acidity, deficiency of essential elements

G – general, all crops; B – brassicas; R – roots; P – potatoes; C – cereals.

	Overall yellowing of the leaf	Death of leaf tips	Leaf margins yellow or brown	Premature death of older leaves	Small lesions on leaves	Yellow mottling between larger veins	Yellow mottling between smaller veins	Red, yellow, orange and purple pigments in leaves	Leaf margins curl upwards	Leaves badly deformed	Weak root system	Brown lesions on stubby roots	Breakdown of internal or growing-point tissue	Plants dwarfed
Nitrogen	G							B			G			G
Phosphorus								R, B						G
Potassium		G	G	G		R, P								
Calcium					C								R, B	
Magnesium					C	R, P		B						
Sulphur	G													
Manganese							R, B, P							
Boron				B						B				
Copper		C									C			C
Molybdenum										B				
Iron	G							R, B			G			
Soil acidity												R, B		G

fluvio-glacial sand, indicates the risk of magnesium deficiency, copper deficiency and potassium deficiency and generally low cation exchange capacity. One may also infer that nitrate fertilizer will be easily leached from a cultivated Podsol and that lime will need to be applied frequently, but in small amounts to avoid trace element deficiencies.

Field observations

The existence of deficiency or toxicity conditions may sometimes be diagnosed through symptoms shown on the leaves, roots or other parts of the plant. Leaf symptoms may take the form of yellowing (sometimes called chlorosis) of various parts of the leaf with persistence of the green colour in other parts. Pigments of other colours – red, orange, purple – may be produced. In some deficiency cases the older leaves may be affected and in others the younger leaves. All colour symptoms may give clues to particular deficiencies or toxicities. The roots of the plant may be short, stubby, brown and 'corky' at the tip, possibly indicating soil acidity. There may be lesions within the storage tissue – a sign that boron deficiency should be considered.

There are several colour guides to nutrient deficiency symptoms which can be consulted. It must be stressed

however that plant symptoms, although very reliable for the diagnosis of problems such as manganese or potassium deficiency, may be very misleading in other cases. For example purple coloration of leaves, described in some texts as an indicator of phosphate deficiency can, as shown in Fig. 14.1, result from several causes that put the plant in danger of death. These include poor drainage, frost, magnesium deficiency, drought or senescence.

Plant symptoms must therefore be interpreted with caution. Also, although they may indicate which element is deficient or present in excess they do not indicate the cause of the condition.

Examination of the soil in the field may give some clues as to physical causes of nutrient deficiency. Rooting patterns should be examined to identify pans, which will physically restrict nutrient uptake. Waterlogged conditions indicate the possibility of denitrification and thus identify a cause of nitrogen deficiency in the plant.

Unless some definite physical cause of a deficiency is established in this way laboratory analysis of soils or plants may be necessary.

Soil and plant sampling and analysis

Both soil and plant analysis can be used either to predict deficiencies and toxicities or to confirm field observations in the event of a crop failing to thrive.

Neither is of much value unless the sample analysed is well taken and is representative of the area from which it is taken.

Sampling plant material

Sampling techniques vary with the crop. In some cases, such as cereal crops, leaf samples are taken. In a few cases, for example suspected boron deficiency in sugar beet or swedes, the root should be sampled.

In any sampling of plant material great care should be taken to avoid soil contamination. This is true when sampling for major element analysis but is critically important when trace element analysis is concerned because the contaminating soil may contain very much more of some trace elements than does the crop and a false result will be obtained.

It is essential to take a representative sample of the crop. One or two whole plants taken from an area may be completely unrepresentative. Samples should be taken from 50–100 plants in a given area. The advisory services, if called in, may wish to take samples themselves, or may advise on which part of the plant to sample.

If a deficiency or toxicity condition is suspected in only a part of a field, 'good' and 'poor' areas should be sampled for comparison in the laboratory. Leaves or other parts of the plant, of the same age or growth stage, should be taken at the same time.

It is extremely important to use containers for transporting the samples which will not contaminate them. Tins, other metal containers and even some paper bags can cause serious contamination of the sample. The best container is a clean polythene bag. After taking the sample indelible labels should be attached, the bag sealed and transferred to the laboratory as quickly as possible.

Soil sampling

Soil sampling is a very skilled job which seldom receives adequate attention. Many of the samples analysed each year give false results because of inadequate sampling by unskilled, untrained and, occasionally, unscrupulous samplers.

Soil is so variable from one place to another within a field and also, in some of its chemical properties, from day to day, that it is very difficult to get an adequate representative sample from an area.

Sampling should be done with an auger with a 20–25 cm long bit of 20–30 mm diameter. Alternatively a corer of the same dimensions can be used but this is not recommended for stony soils. If sampling dry light-textured soils a narrow bladed 'fern' trowel should be used because the soil would fall away from an auger or corer. In this case a small hole should be dug by trowel and a thin slice taken down the side of it.

Whether the samples are being taken for a routine check on lime and major element status, for a more detailed trace element analysis or for comparison of 'good' and 'poor' areas in problem cases, it is necessary to check that the sampling area is reasonably uniform. First, a check should be made of the cropping, liming and fertilizer history of the field. This is particularly important if several fields have been recently made into one. Figure 14.2 illustrates the pH variations in one such field. An exploratory augering of the field should be made to identify variations in soil type.

It is important to avoid sampling an area within two years of liming or four months of fertilizer application. One small particle of lime or fertilizer, inadvertently included in a sample will create a grossly false analytical result. Except for comparative samples, taken in crop failure cases, soil sampling is best done in autumn or early winter, avoiding any obviously unrepresentative parts of the field – headlands, around food and water troughs.

The actual sampling of the selected area is best done on a systematic basis, taking 16–20 auger corings 20–25 cm deep and combining them into one container. Each core of soil is a mini-profile and can yield useful information about texture, colour, drainage and stoniness. Inspection of the subsoil, by re-using the same auger hole to get a core from

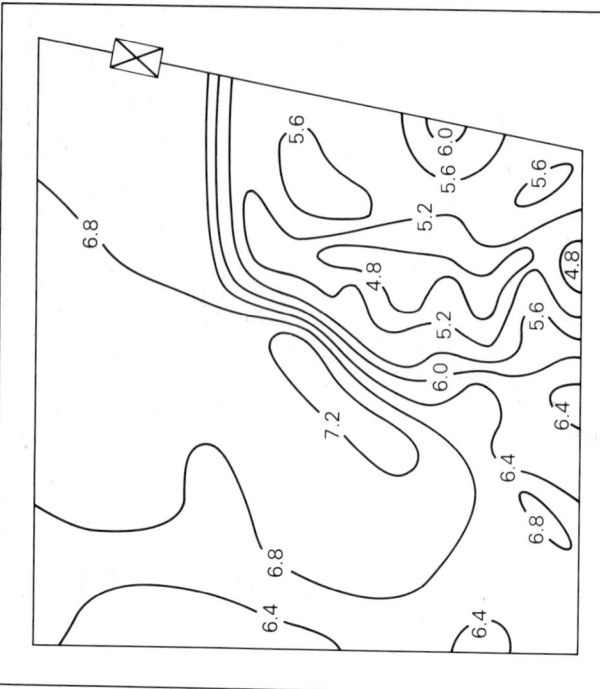

Figure 14.2 pH variations in a field formed by removing hedges from three former fields

25–50 cm below the surface, can be done at every fourth or fifth coring, looking particularly for signs of gleying.

Ideally the area from which a single soil sample is taken should not exceed 1 ha, but this ideal is seldom achieved. In routine sampling, one or two samples per field are all

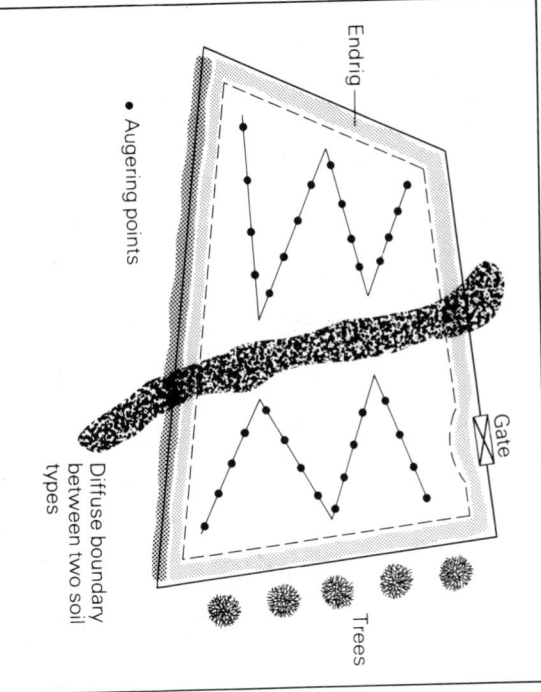

• Augering points

Endrig

Gate

Trees

Diffuse boundary between two soil types

Figure 14.3   A pattern for soil sampling

that can be realistically achieved. Figure 14.3 illustrates a simple sampling plan for a field in which two very different soil types are found with a sharp boundary between them. The zig-zag path followed by the sampler, taking four or five corings on each traverse gives a much more satisfactory coverage than a simple diagonal.

Strong washable linen bags with string closures are ideal containers for the samples. Alternatively, waterproof lined Kraft paper bags may be used. Labelling should be done with waterproof inks. The transport of samples to the laboratory is not so urgent as it is for plant samples.

Laboratory analysis of soils and plants

Both soils and plants can be analysed for the complete range of essential elements. It is customary to determine the *total* amount of an element in plant samples and to express the results as grams or milligrams per kilogram of dry matter. The interpretation of the results is a job for the soil scientist or plant nutritionist because the critical quantities of a nutrient indicating deficiency conditions vary from species to species and also with the age of the plant. Plant analysis in the hands of the expert is very useful for confirmation of a field diagnosis. It can also be used in some cases to predict the risk of deficiency or toxicity in future crops.

The aim of routine soil analysis is to *prevent* problems of acidity and major element deficiencies. It generally involves the determination of soil pH, lime requirement, 'available' phosphate and potassium. Organic matter, available calcium and magnesium may also be determined and there is an increasing need for available sulphur determinations. Determination of the total quantity of the nutrient in the soil is seldom of any value. It is the proportion of the total existing in forms which the plant can take up that is important.

Many analytical techniques have been devised to estimate the so-called 'available' nutrients. None of them can hope to imitate successfully the action of a plant root exploring the soil. There are two basic approaches. One is to measure some part of the total nutrient content which is known to be available to the plant. An example of this is the determination of exchangeable cations. Other methods use extractants such as dilute acids or buffer solutions and relate the amounts of nutrient extracted to field and pot experiment results. If the relationship between analytical result and uptake of the nutrient by field crops is a good one, the method becomes acceptable. More recent methods have employed ion exchange resins and radioisotopes.

However advanced laboratory techniques might become, the key to the value of soil analysis is in the variability of soils in the field and the consequent difficulties in getting a good sample. Because of this, all that can be achieved

from soil analysis is a rough estimate of the likely supply of an element to the plant using terms such as very high, high, moderate, low and very low. It is perhaps, pretentious to use even these five categories and three categories – high, medium and low – would be more realistic.

Rapid testing The hazards of sampling, analysing and interpreting the results are such that any on-the-spot or do-it-yourself methods must be looked upon with grave suspicion. On-the-spot estimation of soil pH, using soil indicators which change colour according to pH or using portable pH meters may be useful on occasions for the diagnosis of liming problems. Also the use of hydrochloric acid (handled with care) to identify 'free lime', in the soil, indicated by a 'fizzy' reaction, can give a very rough idea of lime reserves.

Rapid testing for available nutrients in the field is likely to be of no value whatsoever.

15

Soil acidity and alkalinity

Most of the soils of the British Isles are fundamentally acidic. Exceptions are those soils derived from chalk, limestone, some calcareous tills, sedimentary formations such as Keuper Marl which contain calcium carbonate, and some raised beach deposits.

Because of leaching the soil tends to become more acidic, unless the trend is corrected by liming, and even some older raised beaches are now devoid of calcium carbonate and have become acidic.

Acidity and alkalinity are expressed on the pH scale, low values being acidic and high values alkaline. A pH value of 7.0 is neutral. Figure 15.1 shows the pH range of some types of soil under natural conditions and that of cultivated soils in the British Isles.

Signs of acidity in soils and plants

The degree of acidity in uncultivated hill soils may be diagnosed from the thickness and type of the organic horizons and from the natural or semi-natural vegetation. There is a strong three-way relationship between surface organic matter, pH and vegetation. Table 15.1 gives some examples of this.

The most unreliable indicator species mentioned is moor mat grass (*Nardus stricta*), the presence or absence of which is dictated more by the drainage than by the pH.

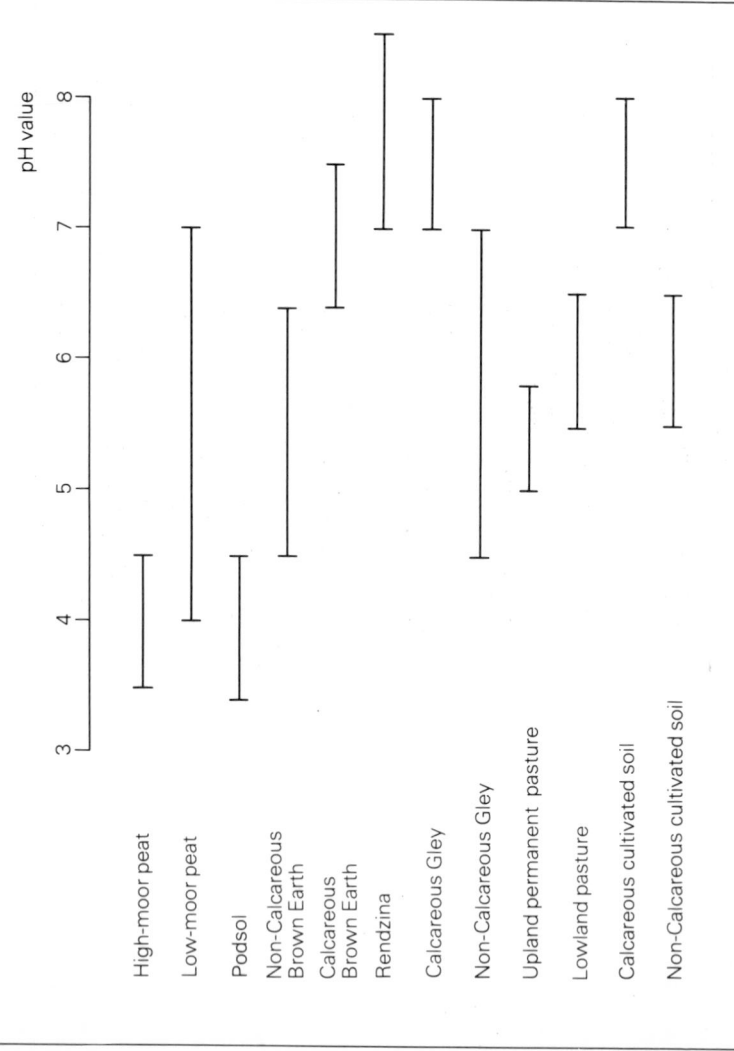

Figure 15.1 The pH range of some natural and cultivated soils

Table 15.1   Surface organic horizons and plant communities as indicators of the degree of acidity in hill and marginal soils

Type and description of organic horizon	pH range	Vegetation
Mull humus. Organic matter completely incorporated into surface mineral horizons. Earthworms abundant	Greater than 5.5	Good pasture with bent (*Agrostis* spp.), fescues (*Festuca* spp.) and wild white clover (*Trifolium repens*)
Mull/moder humus. Most organic matter incorporated but a slight build up (1–2 cm) of black or dark brown humus at surface. Some earthworms present	5.2–5.5	Poor pasture with little white clover, sheep's fescue (*Festuca ovina*), a little moor mat grass (*Nardus stricta*). Small herbs such as tormentil, heath bedstraw
Moder/mor humus. Some organic matter incorporation. Discrete black humus horizon. Slight build-up (less than 1 cm) of decomposing litter at surface. Earthworms absent or few in number	4.5–5.2	Sheep's fescue with some heather (*Calluna vulgaris*) and blueberry (*Vaccinium myrtillus*), bell heather (*Erica cinerea*), moor mat grass. No wild white clover
Mor humus. Distinct black humus horizon. overlain by discrete horizons of decomposing and undecomposed litter. No earthworms	3.5–4.5	Heather, with some bell heather, blueberry, cross-leaved heath (*Erica tetralix*), some mosses
Black amorphous peat	4.5–5.0	Purple bent (*Molinia caerulea*), some moor mat grass. Cross-leaved heath, polytrichum moss, sphagnum moss
Brown fibrous peat	3.5–4.5	Heather with red sphagnum moss as a bottom layer and some cotton grass (*Eriophorum vaginatum*)
Pale brown or yellow peat	3.0–3.5	As above but sphagnum moss becomes dominant with cotton grass and heather

There is no way of diagnosing the degree of soil acidity in soil under arable crops or temporary grass from the appearance of the soil. If clovers are present in temporary grass swards it is a sign that the pH is above 5.0. It should not, however, be concluded from the absence of clover that the soil pH is less than 5.0. There are one or two weed species, such as spurrey (*Spergula arvensis*) and sheep's sorrel (*Rumex acetosella*) which are regarded as acidity indicators. Their presence is by no means conclusive evidence of acidity but it is worth checking the pH to find out. In these days of intensive weed control it is virtually impossible to diagnose soil acidity from the weed flora.

Symptoms in crops

Symptoms on the leaves and roots of crops may be used to diagnose acidity. No symptom is absolutely reliable and soil pH should be determined to confirm the diagnosis. The absence of plant symptoms does *not* rule out acidity.

Of the cereal crops rye and oats are most tolerant of acidity and seldom show symptoms. Wheat and barley plants are dwarfed and develop thin 'spiky' leaves with yellow tips turning brown. Reliable root symptoms are also found. The roots are thick and 'stubby', ending abruptly, branching little and having extensive brown areas on their surfaces.

Root crops and brassicas become stunted and many plants die out completely, leaving areas of bare soil. The leaves tend to 'cup', the margins turning upwards. Yellowing and browning of leaf margins sometimes occurs. Brassica crops sometimes show rich orange, yellow and purple colours but these symptoms are not specific to acidity. Potatoes are tolerant of acidity and seldom show symptoms.

Clovers and lucerne may fail to produce root nodules, and there will be brown lesions on the weak root system. Plants will wilt easily in dry periods. There are no specific symptoms of acidity in grasses. In grass/clover swards acidity manifests itself by the disappearance of clover and the gradual takeover of acid-tolerant grasses such as Yorkshire fog (*Holcus lanatus*) and red fescue (*Festuca rubra*) from sown acid-intolerant species such as ryegrasses and cocksfoot. There is insufficient time for this to happen in short-term grass leys but permanent grassland is affected as acidity develops.

Patterns of acidity in the field

It is very seldom that the whole crop in a field will fail because of acidity. There is usually an irregular pattern of failure, partial failure and reasonably health growth. The margin between the three in terms of soil pH can be very small indeed, particularly in sensitive crops such as sugar beet in which complete failure may occur within a metre or so of a satisfactory crop.

The patchiness of acidity in fields is usually a result of soil variability. Soil in slightly sandier parts of the field may be leached of lime more rapidly than other parts. Sometimes soil on the crests of small ridges or hillocks is affected before lower lying parts of the field. When lime is applied at a standard rate it is impossible to allow for such soil variations and there may be differences of more than a unit of the pH scale within a field (Fig. 14.2) but routine pH determinations on a sample of the whole field will give an average value. The key to the prevention of patchy failure caused by acidity is to maintain the pH, at all times in the optimum range (Fig. 15.2) well above the minimum level required for a proposed range of crops, at the same time avoiding overliming.

Regular patterns of crop failure, in strips across a field, are usually caused by poor distribution of lime. Some alarming patterns can be seen immediately following application, before cultivations obscure the lines of concentration behind the distributor.

Tolerance of acidity by crops

Figure 15.2 shows the 'optimum' pH range for crops grown on mineral soils in the British Isles and the range over

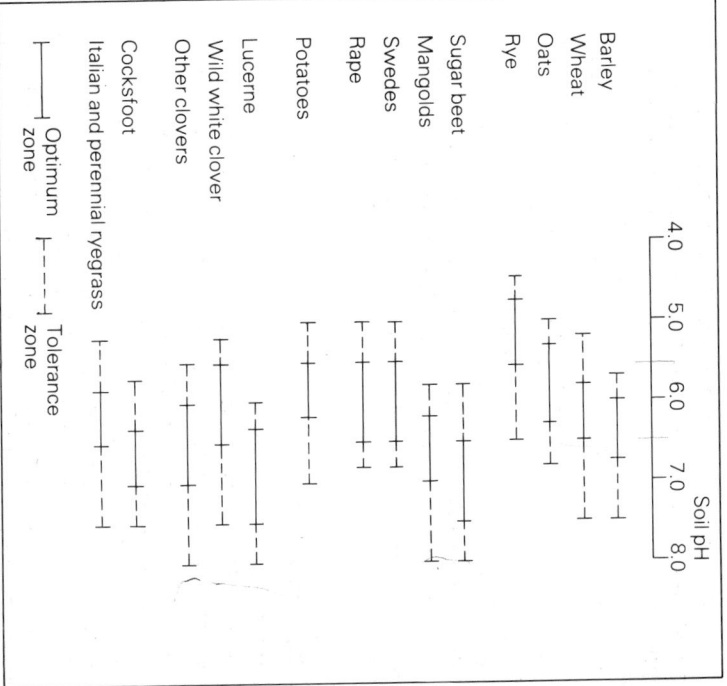

Figure 15.2   Acidity tolerance of the main crop species

which the crops will usually grow satisfactorily. It is *not* recommended that the extremes of the tolerance range should be used. In the optimum range the crop will suffer no acidity problems.

Causes of injury to crops

The causal factors in the injury or death of plants growing in acid soils are not simple. The following points are worthy of note:

- In mineral soils, as acidity develops, the soil solution contains increasing amounts of manganese and aluminium. These elements can be taken up by the plant in sufficient quantities to be toxic. This is one of the most frequent causes of acidity damage.
- In strongly acidic soils there is very little conversion of organic nitrogen to the ammonium and nitrate forms which are available to the plant.
- Phosphorus, a major plant nutrient, is very much less available to plants in very acid soils (pH less than 5.5) than in slightly acid soil (pH 6.0–6.5).
- The trace element molybdenum may be deficient in acid soils. Also, in Podsols leaching of trace elements such as copper and zinc has occurred for thousands of years and the total content of these elements may be low.

- Failure of root systems to penetrate large volumes of soil is a major factor.
- Certain disease organisms, such as *Plasmodiophora brassicae*, which causes 'finger and toe'in swedes and 'clubroot' in other brassica crops, thrive under acid conditions and, once established, are not easily eliminated by liming.
- The cause of injury or death is seldom calcium deficiency, although it is a contributory factor. Any of these factors can be involved in the failure of crops to thrive in acid soils. For example a plant may die of a combination of manganese toxicity, phosphate deficiency and nitrogen deficiency, all having arisen because of soil acidity.

Acidity and fertilizer application

There are two ways in which fertilizer use can create soil acidity problems.

Exchange acidity   When a soil which has not been limed for some time begins to become acidic, hydrogen ions attach themselves to the cation exchange complex. They can be displaced from it by other ions and released into the soil solution. This phenomenon is known as exchange acidity. As an example, a soil may have a pH value in water suspension of 6.0 but after adding a neutral salt such as potassium chloride the meter may give a reading of 5.5. A similar effect occurs in the field after the addition of potassium chloride as a fertilizer. This can give rise to loss of yield through acidity in newly fertilized crops. Root scorch occurs and the leaves of cereals may take on a striped appearance of yellow and green and die from the tip. Affected plants contain large quantities of manganese, aluminium and, commonly, increased amounts of other elements such as copper, nickel and potassium. Injury to plants is the result of multiple toxicity.

Effects of some nitrogenous fertilizers Any fertilizer containing ammonium salts brings about some acidification of the soil. The commonest nitrogen source in fertilizers, in the British Isles, is ammonium nitrate. This substance, particularly when applied at high rates (200–400 kg N per hectare), will cause considerable acidification. The pH of any soil should be checked regularly, preferably every year, in these circumstances.

Some 'straight' nitrogen fertilizers containing ammonium nitrate also contain calcium carbonate. This will, to some extent, neutralize the acidity caused by the ammonium nitrate component. Examples of this are 'Nitrochalk' and 'Nitra-shell'. Ammonium nitrate fertilizers such as 'Nitram', which do not contain calcium carbonate, will have the full acidifying effect.

Treatment of soil acidity with lime

The low pH of an acid soil can be raised by adding lime. The term 'lime' includes calcium carbonate, magnesium carbonate, and related compounds such as calcium oxide, CaO, and calcium hydroxide, Ca(OH)$_2$. By far the most commonly used liming materials in the British Isles are chalk which consists mainly of calcium carbonate and limestones composed of calcium and magnesium carbonates.

Materials formerly used quite widely, such as burnt lime, CaO, slaked lime, Ca(OH)$_2$, and waste limes from sugar beet factories are now seldom used. They have no advantages over chalk or limestone and they are unpleasant to use.

The specifications of liming materials are legally controlled by the Fertilizer and Feeding Stuffs Act which requires the producer or seller to quote the quality of the lime in terms of calcium oxide or calcium carbonate equivalent. The fineness of grinding is also controlled. All the material must be smaller than some 3 mm and 40 per cent of it must be small enough to pass through a standard 100-mesh sieve, thus ensuring a high proportion of very fine, quick-acting material.

Liming policy

Decisions required concern the type of lime, the amount of

lime, the frequency and time of liming. Inadequate liming can have catastrophic effects on crops. None the less, in times of financial stress, liming is one of the first essential operations to be neglected, perhaps because the effects of neglect are not obvious for a year or two. Strong evidence of this comes from the liming statistics for 1970–80. As the decade progressed less lime was applied and, at present, insufficient lime is being applied (on a United Kingdom basis) to maintain the pH levels in agricultural soils. This has resulted partly from the removal of subsidies and partly from general economic conditions in farming.

The difficulties of maintaining the soil in a pH range which is optimal for a multi-crop farming system are shown all too clearly in Fig. 15.2.

A farmer growing barley and potatoes will find difficulty in maintaining the soil pH within the narrow pH band that is optimal for both crops. If sugar beet is added to the cropping sequence, the task becomes impossible and a compromise is essential. In practice, most farmers would adopt the higher pH range suitable for sugar beet and barley and would risk potato diseases, such as common scab, that might result.

Type of lime

The farmer's choice of type of lime is often limited by the geographical distribution of chalk and limestone deposits.

Liming contractors will buy from the cheapest source of good material and transport costs usually dictate that local chalk or limestone will be offered to the farmer. Most materials on offer, irrespective of type, are similar in neutralizing value, equivalent to 85–95 per cent calcium carbonate.

Most areas in the British Isles are close to a source of lime which is currently being worked. In areas where there is a choice of sources there is a slight advantage in the speed of action of chalk and softer limestones, such as Oolite, over the harder carboniferous limestones, but this does not justify paying a premium for them.

The one case in which the payment of a premium *is* well worth considering is that of magnesian limestone. Many areas of the British Isles are deficient in magnesium, a major plant nutrient. The serious and widespread occurrence of hypomagnesaemic tetany in ruminant livestock, especially cattle, in the 1940s and 1950s pointed up this problem. The regular use of magnesian limestone, either for every application or for alternate applications will help to alleviate magnesium deficiency in both crops and stock. It is important to stress, however, that although liming with magnesian limestone will usually prevent symptoms of magnesium deficiency in crops it cannot, alone, be relied upon to prevent problems in stock. Dosing of stock with magnesium compounds is often the only effective remedy for hypomagnesaemia.

The Fertilizer and Feeding Stuffs Act defines any limestone containing more than 3 per cent magnesium (11 per cent magnesium carbonate) as magnesian. The magnesian limestone of north-east England generally contains much more magnesium than this. A good sample would contain 50 per cent calcium carbonate and 40 per cent magnesium carbonate plus impurities. This material is roughly equivalent in neutralizing value to most calciferous limestones.

There are, on the market under trade names, some products similar in analysis to limestones but claiming to contain more of certain trace elements. These products are sold at several times the price of ground limestone. There is no justification for paying such a price.

Amount of lime

The quantity of lime required to raise the pH of soils to various values can be assessed fairly accurately by laboratory methods. The advisory services will analyse soil samples for pH and lime requirement, as will some fertilizer and lime merchants. Each soil has a specific lime requirement to bring it to a given pH value which will depend on texture, type of clay, and type and quantity of organic matter. For soils with a particular pH value the lime requirement of low organic matter sands will be lowest and that of high organic matter clays and peats will be highest.

The range of lime requirements varies from nil for soils with pH values of 7.0 or more to 20 t/ha of calcium carbonate for soils with pH values of less than 4.0. For most cultivated soils, on which routine liming is practised, the lime requirements will be in the range 4–8 t/ha. Application of ground limestone at rates greater than 8 t/ha is both wasteful and dangerous, because excessive amounts of lime will be leached and dangerous, because of the risk of inducing trace element deficiencies.

Frequency and time of liming

The intensive liming campaign of 1938–45, followed by high liming subsidies for a period of 30 years has done much to reduce soil acidity. In United Kingdom soils, except in land reclamation cases, the need is for maintenance liming.

The frequency and timing of lime application are critically important. In most circumstances, liming will be required at intervals of four to six years. The following factors influence frequency:

● In a given rainfall area, sandy soil will become acid more rapidly than clay soil due to the low cation exchange capacity and the more rapid percolation of water through the soil.

- Soil in low rainfall areas will acidify more slowly than that in high rainfall areas.
- Soil which is being intensively cropped, with the aid of high rates of ammonium-nitrogen fertilizers, will become acid more rapidly than similar soil cropped less intensively.
- The soil pH should be checked four years after liming and a decision made on renewed application at that stage. It is very important to apply lime timeously. The tendency is to apply lime during the winter or spring immediately before the most demanding crop is grown. This is too late for the lime to be fully effective for that crop. Lime should be applied fully one year before this crop is sown – that is not in the previous winter, but in the one before that.

Method of application

Surface application of lime is the most satisfactory method. Any machine which will give a reasonably uniform distribution across the field may be used. It is seldom necessary, except in land reclamation to use split dressings. The first ploughing after lime application will ensure a reasonable vertical distribution of the lime. In soil where minimum tillage is practised the lime will incorporate slowly from the surface. This is desirable because in these conditions it is the surface 5–10 cm which become acidic

most rapidly and there is a need for lime to be concentrated at the surface.

Reclamation of acid soils

Most areas of lowland soil were claimed for agriculture many years ago, but during the Second World War there was a movement to bring more and more marginal and hill soil into cultivation. There are still many areas of potentially good soil in Britain, lying between the 250 m and 350 m contours, which can be reclaimed and worked under a rotation such as three years' cropping, six years' grass. If these soils are Podsols, under heather, or Acid Brown Earths under grass heath, the lime requirement is very high and can amount to 12–20 t of calcium carbonate per hectare. When reclaiming such soil it is foolish to try to satisfy this lime requirement in one application. To do so leads to problems such as trace element deficiencies. It is also inefficient in that much of the lime would be lost by leaching. There is no standard procedure for reclamation of acid hill soils but the following is given as an example, using a Podsol, pH 4.5, with no iron pan and having a cover of pure heather.

The indigenous vegetation should be destroyed by flailing, or some other vigorous action to break up the living material. Burning may be necessary in the case of heather. Ground limestone (5 t/ha) should be applied to the surface of the soil before ploughing. Deep ploughing or ploughing with subsoiling is often needed on Podsols in order to break up iron pans. A further 5 t/ha of ground limestone should be applied to the soil after ploughing and incorporated by vigorous cultivation.

If a tilth can be obtained the soil is then ready for a pioneer crop of rape or oats to be sown, but the soil will almost certainly be very deficient in phosphate. A dressing of 500–700 kg/ha of ground mineral phosphate or its equivalent of another phosphatic fertilizer should be applied in addition to a standard fertilizer for the crop to be grown. High yields cannot be expected in the first crop or even the second crop. The soil has been establishing an acid regime sometimes for centuries and the chemical reactions required to change this regime take place over months or years. For example, problems of exchange acidity will be very likely to occur. The first crop to be planted should therefore be an acid-tolerant crop such as rape or oats and only after two or three years should crops such as barley which do not tolerate acidity be attempted.

A major problem associated with reclamation of acid soil is nitrogen deficiency in the first crop. The reason for this is that the liming programme triggers off a very vigorous decomposition of organic matter. The carbon/nitrogen ratio of this organic matter is very high and a high input of nitrogen is required by the micro-organisms. It is often not economical to apply large quantities of nitrogen to the pioneer crop and it is also very difficult to judge the exact amount required so that it will be necessary to accept that areas showing nitrogen deficiency symptoms will be found.

Another major problem which occurs in the reclamation of acid soil, particularly Podsols and Peats, is deficiency of copper in the crop. Copper, in common with other trace elements, has been leached from podsolic soils and if the parent material had a fairly low content of copper in the first instance, the total quantity following podsolization will be very small. This problem is exacerbated by the fact that the chemical reactions following liming will reduce the availability to the plant of what copper is present and can reduce cereal crop yields severely.

Most reclaimed hill or marginal soils are returned to grass after two or three years' cropping. A further dressing of about 4 t of ground limestone per hectare should be applied before the grass is sown and it is preferable to establish the grass under a cereal crop. If the reclamation has been handled satisfactorily a good grass/clover sward should now be easy to establish using high-yielding grasses such as ryegrass.

Liming practice in the future

There are, at present, few departures from traditional

liming practice, which has proved its worth over a 50 year period. There are, however, two possibly complementary developments which may occur.

The use of lime as a fertilizer

Soil acidification can occur quite rapidly under intensive crop production systems. Also surface soil under arable crops tends to become acid where no cultivations are used or under minimum tillage systems. In these circumstances and even where cropping systems are less intensive, there is a case for re-thinking traditional liming practices.

The availability of fertilizer distributors which give accurate and uniform distribution has brought annual lime application within the capability of many farmers. Based upon estimated losses of lime in crops and drainage water an annual schedule of liming could be established by the farmer and his adviser. Annual application of 0.5–1.2 t/ha would be required and systems could be tailor-made to cropping requirements. It certainly should not be beyond the ingenuity of the agricultural engineer to ensure that spreading could be achieved with minimum inconvenience.

The use of coarsely ground limestone

The above system could be operated within the terms of the present Fertilizer and Feeding Stuffs Acts. Another development would require amendment of the Acts which might be a long and tedious procedure. The suggestion is that the present legislation on fineness of grinding which requires 40 per cent of the limestone to pass a 100-mesh sieve might be relaxed.

The law was undoubtedly justified when it was made some 50 years ago. At that time very large areas of soil were so acid that barley and sugar beet growing was severely restricted. Crop failures caused by acidity were frequent and the need was to raise the lime level of soil throughout the country rapidly – a need exacerbated by the Second World War.

At present, maintenance liming is often all that is necessary and this could probably be achieved by using 1–3 mm ground limestone, possibly with a 10–20 per cent inclusion of material finer than 1 mm in size. On a national scale the saving of energy used to grind to present specifications would be great. The future use of coarsely ground limestone applied by fertilizer distributor on an annual basis is certainly feasible even with the equipment at present available.

Calcareous soils

Naturally alkaline soils in the British Isles are dominated by calcium carbonate and, to a lesser extent, magnesium carbonate with none of the major problems associated with the much more alkaline sodium-saturated soils of arid parts of the world. The pH of calcareous soils is usually between 7.0 and 8.0.

Naturally Calcareous soils in the British Isles are derived mostly from chalk, from calciferous limestones of varying ages, usually containing 90 per cent or more of calcium carbonate, or from magnesian limestone. There are also, around the coast, raised-beach deposits of shell sand.

The only other major group of Calcareous soils is the Calcareous tills such as Chalky Boulder Clay and clays derived from some sedimentary formations such as the Lower Lias. These are usually well-structured fertile soils and, if well managed, present few problems.

The main physical problems associated with chalk or limestone soils are a tendency to dryness and shallowness. The parent rock is commonly near the surface and because drainage is free, water availability is restricted. Many of these soils contain high proportions of silt-sized calcium carbonate but, in contrast with other silty soils the surface structure is stable, relatively free from capping and compaction and recovers rapidly from maltreatment by farm machinery.

Chemical problems in Calcareous soils are mainly caused by the great excess of calcium and magnesium over other elements. As a result the cation exchange complex is calcium dominated, with magnesium sub-dominant and all

other cations including potassium and ammonium applied in fertilizers are subject to leaching if heavy rain occurs. The high pH also helps to induce deficiencies of manganese, iron, boron and copper. Phosphate fixation is a major problem. These deficiencies and remedies for them are described in detail, for each element, in Chapter 16.

Overliming

Some of the problems encountered in Calcareous soils can be induced in acid soils by overliming. Manganese, copper and boron deficiencies may all be induced on certain soils in this way.

Deliberate overliming based on the fallacy that 'double the application will do twice as much good' is now rare, although it was quite common in the days of high lime subsidies.

More frequently overliming occurs by accident or bad application. At the time when burnt lime, hydrated lime and waste limes were widely used, the materials were carted on to the field and left in heaps before distribution. Large lumps of lime, particularly wet waste limes which were impossible to distribute evenly, gave rise to serious patchy overliming and it was not unusual to see manganese deficiency caused by overliming, and manganese toxicity caused by acidity, within the same field and sometimes within a few metres.

There is really no excuse for uneven distribution of ground limestone but some contractors leave badly 'striped' fields, and the unevenness cannot be eliminated by cross-cultivation.

The only preventive measures that can be used for overliming are careful assessment of amounts of lime required and even more careful choice of liming contractor. If overliming has occurred, only leaching over a period of many years will correct it. Meantime preventive or corrective treatments for trace element deficiencies may be required.

An extreme case of overliming, with all its problems occurred on a sugar beet crop grown by a farmer who had inherited the problem from his 'lime-happy' predecessor. The unfortunate combination of a high organic matter sandy Podsol reclaimed 15 years before, into which cheap lime had been poured from an adjacent lime quarry for a period of 10 years, and a dry season had induced classical symptoms of both boron and manganese deficiency. The overliming occurred between 1940 and 1950. The amount of 'free lime' in the soil in 1950 was equivalent to 35 t/ha of limestone and routine spraying of barley for manganese deficiency was still necessary 25 years later. Admittedly, this was an extreme case but the after-effects of overliming commonly last for 10–15 years.

Plant nutrient problems

16

It is beyond the scope of this book to give detailed fertilizer recommendations for individual crops. The advisory services and reputable fertilizer manufacturers and distributors will supply them, and will give advice on request. It is possible to give some guidance here on treatments for trace element deficiencies and this is done in Table 16.1 (page 213) but it would be wise to consult local advisers before treating any suspect deficiency, both for confirmation of diagnosis and for advice on remedial treatments.

Deficiencies of the major elements nitrogen, phosphorus, potassium and calcium are not now widespread in the British Isles, thanks to constructive fertilizer and liming policies backed up by a highly efficient fertilizer industry. For many reasons, including the greatly increased intensity of agriculture with its greater demands on available nutrients from the soil, deficiencies of other major and trace elements have increased and are still increasing.

At the other end of the scale toxic quantities of some elements can cause plant damage or death either as a result of high indigenous levels of the nutrient or as a result of pollution. Again, for detailed information about pollution other sources must be consulted.

Nitrogen

Nitrogen deficiency can occur on almost any type of soil, but is most likely in cool, wet areas either on acidic, freely drained soils or on soils that are frequently waterlogged.

Causes of nitrogen deficiency

There are many causes of nitrogen deficiency. Some are directly associated with a lack of available ammonium (NH_4^+) and nitrate (NO_3^-) in the soil:

Leaching is most severe in low-organic, light-textured soils in high rainfall areas, and causes loss of available nitrogen.

Denitrification is found especially in Gley soils where alternate oxidizing and reducing conditions occur and causes conversion of nitrate to nitrogen.

Acidity causes enhanced leaching of ammonium-nitrogen and prevention of mineralization of organic nitrogen compounds.

High carbon/nitrogen ratio materials, such as straw, incorporated in the soil, cause bacteria to compete with the plant for available nitrogen.

Inadequate or badly timed fertilizer applications, especially if available nitrogen from within the soil is low, can bring about early-season nitrogen deficiency.

Nitrogen deficiency can also be caused indirectly, by biological or physical phenomena, even though the soil is well supplied with available nitrogen:

Pan formation prevents roots from tapping available nitrogen in the subsoil.

Cold, wet conditions, especially at the time of seedling emergence, cause slow penetration of the soil by roots and thus restrict the amount of nitrogen taken up. This is often a temporary effect.

Poor drainage also restricts root penetration.

Drought, as well as restricting roots, also prevents diffusion of nitrate towards the roots.

Root damaging organisms, such as eelworms and some fungi, render roots inefficient and incapable of taking in available nitrogen.

Diagnosis of nitrogen deficiency and its causes

Early and correct diagnosis of nitrogen deficiency and its cause can be very rewarding.

Plants deficient in nitrogen will be pale green, yellow green or yellow. Usually older leaves are affected first, turning yellow and dying from the tip. The essential feature is an overall yellowing of the leaf, veins and all, brought about by lack of chlorophyll. The leaf symptoms are fairly reliable indicators of nitrogen deficiency, but do not indicate the cause.

Soil analysis for either 'total' or 'available' nitrogen helps very little in either prediction or diagnosis of nitrogen deficiency. This is very unfortunate considering that nitrogen influences crop yields more than any other single plant nutrient. The problem is that day-to-day as well as seasonal variations in amounts of available nitrogen in the soil are very large. Further, there is no way of assessing, with any certainty, how much organic nitrogen will become available in a particular season. Taking this into account it is important not to attempt to assess the nitrogen status of soils or plants by 'do-it-yourself' methods.

The identification of the cause of nitrogen deficiency in a particular case can call for the combined efforts of the farmer, agricultural adviser, soil chemist, soil physicist, mycologist and nematologist.

Prevention and correction

Adequate applications of fertilizers and manures as recommended by the advisory services will prevent most cases. Nitrogen fertilizers must be applied to meet only the needs of the current crop. Excessive applications must be avoided. Residual effects for the next crop are usually small and cannot be relied upon. The growing of clovers and other legumes will help considerably in the prevention of nitrogen deficiency.

Success or failure of corrective measures depends upon accurate diagnosis of the cause of the deficiency. If it results from inadequate fertilizer application, leaching of available nitrogen or ploughing in of high carbon/nitrogen ratio material, very rapid responses will occur within days to a top dressing of nitrogenous fertilizer. In cases of drought, one can only wait for rain, or, if the apparatus is available, irrigate.

Yellowing of cereal plants at the two to three leaf stage can occur as a result of cold weather, even though nitrogenous fertilizer has been applied in the seedbed. Impulsive top dressing of such crops with more 'nitrogen' should be avoided as it can lead to overapplication resulting, for example, in lodging of cereal crops and susceptibility to disease in many crops. It is wise to wait for a week or two. The crop will usually grow away as the soil temperature rises. If not, there is probably another reason for the deficiency.

If nitrogen deficiency results from acidity there will be other symptoms on the crop apart from yellowness ('spiky' leaves, stubby roots) and the prime need is to check the pH and to correct any acidity with an urgent application of lime.

Crops affected by nitrogen deficiency from indirect causes will sometimes respond to top dressing but usually the results are disappointing. In cases caused by pans or inadequate underdrainage top dressing will sometimes help by stimulating a surface root system which will give the plants a chance to dry out the top few centimetres of soil by transpiring more water. This process can be cumulative and will create a more satisfactory living space for roots. It must be stressed that this is often only a rescue operation and the causal factor will still require attention.

The addition of extra fertilizer nitrogen can also alleviate conditions caused by eelworms or root fungi although the final crop yields will usually be disappointing. Brassica crops suffering from club-root can respond well to fertilizer nitrogen and cereal root systems attacked by eelworms can be partially revived, but again the causal factors must be treated before the next crops.

Nitrogen fertilizers for top dressing   There is conflicting evidence and consequently much nonsense has been talked and written about the pros and cons of ammonium and nitrate nitrogen. If symptoms of nitrogen deficiency occur and the cause is direct, i.e. lack of available nitrogen, the problem is one of urgency and the most convenient 'straight' nitrogen fertilizer should be used. Most of the solid fertilizers are now based on ammonium nitrate, either concentrated (35 per cent nitrogen) or at lower concentrations (15–25 per cent nitrogen).

Effects of excess nitrogen

Excess of nitrogen for crops usually results from overapplication of fertilizer. Plants containing excess nitrogen take on a dark blue-green colour. Leaves are soft, luxuriant and fleshy, and stems are weakened. Lodging in cereals crop is one effect, the stems breaking under rain and wind long before ripening, resulting in a 'flat' crop with ripening and harvest problems. This was a very serious problem 10 to 20 years ago before the breeding of stronger strawed cereals which resist lodging to some extent. Other crops, potatoes for example, are susceptible to mechanical damage by machinery or wind which creates sites for disease infection. Potatoes and root crops containing excess nitrogen have low dry matter contents and poor keeping quality.

The farmer has no satisfactory way open to him to correct excess nitrogen, although in crops like potatoes, plants can be 'toughened up' by adding some extra potassium fertilizer. Excess nitrogen application must be avoided by careful reference to the recommendations made by the advisory services or reputable fertilizer firms.

There is strong evidence that some of the excess nitrogen taken up by the plant is not converted to protein but remains as non-protein nitrogen. This is not only inefficient use of nitrogen by the plant but leads on to inefficient use by the animal and the risk of ill-effects on human beings eating the plants.

Excess nitrate nitrogen in the soil is subject to leaching into drains and water courses and can be a major pollutant in lakes.

Phosphorus

Many of the soils of the British Isles are naturally deficient in phosphorus. During the last century this has been alleviated in most cultivated and well-managed grassland soils because of the residual effects of large applications of phosphate fertilizers. Phosphate deficiency is most serious in very acid soils (pH less than 5.0) and in Calcareous soils (pH greater than 7.0). Deficiency is more likely to occur in freely drained soils than in comparable poorly drained soils.

Causes of phosphorus deficiency

The main cause of phosphorus deficiency in crops is the fixation of phosphates in unavailable forms by iron and aluminium in acid soils and by calcium in Calcareous soils.

Diagnosis of phosphorus deficiency

There is no reliable plant symptom of phosphorus deficiency. The frequently quoted 'purpling of the leaves' is certainly not specific to phosphorus deficiency. Plants suffering from phosphate deficiency are dwarfed but seem to be capable of survival so that, in extreme cases, a full population of tiny plants can be found, the yield from which is very low indeed.

Soil analysis on a routine basis, if undertaken by a reputable advisory agency, is a reliable guide to the risk of phosphorus deficiency. It should be regarded as a tool in prevention of deficiency rather than in correction although both soil and leaf analysis can be used to confirm a deficiency.

Prevention and correction

The first priority in prevention is to raise the pH of the soil, if very acid, to the middle range 5.5–6.5 according to the proposed cropping. Nothing can be done to lower the pH of Calcareous soils and phosphorus deficiency must be accepted as an inherent hazard in these soils and fertilizer programmes planned accordingly.

Regular application of recommended rates of phosphate fertilizer and farmyard manures for crops over a period of 20 to 30 years, with supplementary dressings for soils found deficient by soil analysis, will gradually reduce the risk. The reason for this is the build-up of residual phosphate from each application. Only 25 per cent or less of the fertilizer phosphate applied to a crop is taken up, and a build-up of total phosphate, some of which is available to future crops, is inevitable. Extra phosphate should be added over a period of years to soils that are found on analysis to be deficient. This will supplement phosphates applied in routine fertilizers. Supplements can be applied in forms which are cheaper than the water-soluble phosphates used in compound fertilizers. The usual supplements become available in the soil over a period of some years and may be applied at intervals of three to six years. Supplementation should eventually become unnecessary.

The old stand-by for this purpose was basic slag from the steel industry but this has now virtually disappeared from the market. Supplements now available include ground mineral phosphate (GMP) which is most effective on acid soils (pH less than 5.5). There are also various proprietary mixtures of GMP and water-soluble phosphate which will be most effective on mildly acid or alkaline soils (pH greater than 5.5). The advisory services should be consulted about the frequency and amounts of supplements required in individual cases.

If a deficiency of phosphorus is diagnosed in a growing crop, an urgent application of water-soluble phosphate fertilizer should be made to supply 50–80 kg P_2O_5 per hectare. If phosphate deficiency is the sole cause of the problem the response in the crop will be immediate and vigorous. If acidity is also involved, response to the applied phosphate will be poor and immediate attention must be given to liming.

Acute phosphorus deficiency is now restricted mainly to hill and marginal areas.

Excess phosphorus

Applications of excessive amounts of phosphate fertilizer are wasteful and uneconomic but recorded cases of phosphorus toxicity in crops are rare in the British Isles. The pollution of watercourses by phosphate is often blamed on fertilizer application. Actually this is a most unlikely source of pollution because of the capacity of soils to immobilize and fix phosphate.

Potassium

Potassium deficiency occurs most commonly on Podsols, deep Peats and Calcareous soils. The main causes of potassium deficiency are excessive leaching in acidic or formerly acidic soils, lack of potassium in the parent material and excess of calcium either present naturally in the soil or resulting from overliming.

It can also be caused in intensive cropping systems if high levels of fertilizer nitrogen are used with low levels of potassium. It is unlikely to happen in crops which have received a liberal dressing of farmyard manure or slurry because this will supply large quantities of potassium.

Diagnosis of potassium deficiency

Leaf symptoms are the best indicators of potassium deficiency. Figure 16.1 illustrates the symptoms on a potato leaflet. They are both striking and reliable and similar symptoms may be found on sugar beet, swedes, top fruit and soft fruit crops. They occur typically on the older leaves of the plant because the plant is capable of withdrawing potassium from these leaves to young, actively growing leaves when there is a shortage.

The first symptom is yellowing of the tip and leaf margin. These parts then die as yellowing followed by browning spreads inward, between the major veins. The whole leaf will then die or the dead parts between the veins will dry out and be eroded away.

Clovers show rather different symptoms. Numerous black or dark brown spots are found along with marginal

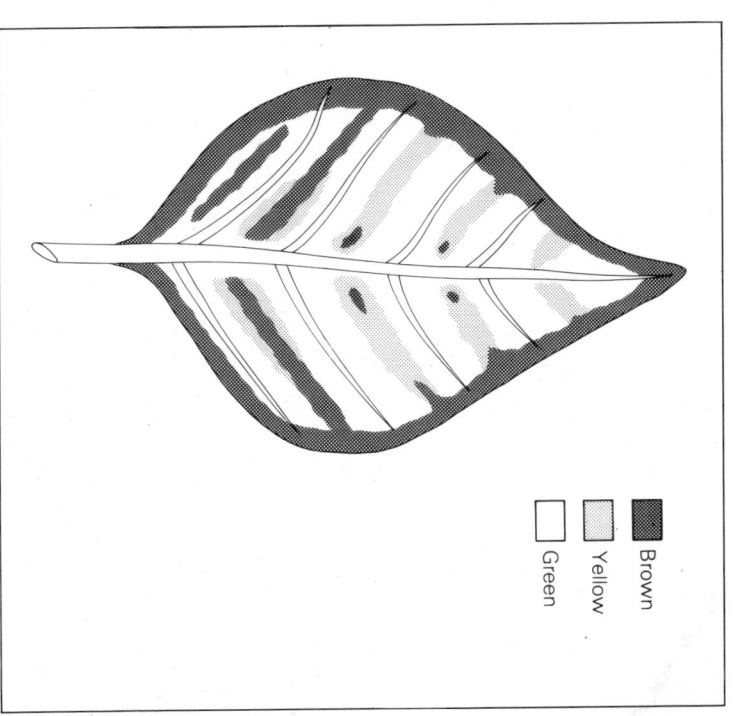

Figure 16.1   Symptoms of potassium deficiency in potato

Brown

Yellow

Green

browning. Symptoms are relatively rare in cereal crops, which require relatively little potassium.

Soil analysis is a reasonable guide to the risk of deficiency but some soils with high natural reserves of potassium have a remarkable capacity to recover after their available potassium has been depleted and analysis of a sample taken in the autumn may underestimate the potassium-supplying power of the soil in the following summer. Leaf analysis may be used to confirm field diagnosis.

Prevention and correction

It is neither easy nor desirable to build up reserves of soil potassium by fertilizer application. Excess may well be taken up by the plant or leached from the soil. It is better to pay careful attention to the potassium requirement of individual crops. Farmyard manure and slurry are valuable sources of potassium and, if they are used, the rate of fertilizer potassium can be reduced.

If a crop develops potassium deficiency symptoms late in the season, when the crop is maturing, they may be ignored. The leaves on which the symptoms occur have served their purpose and are dying. Early season symptoms require urgent treatment with a top dressing of potassium fertilizer. Such symptoms are not now common

in field crops, thanks to regular fertilizer use, but they can occur in the clover in grass/clover leys especially if the sward is cut and not grazed. They may be corrected rapidly by applications of potassium, sometimes with quite startling effects on the proportion of clover in the sward. If this occurs there is a considerable risk of animals grazing the sward suffering from 'bloat', a condition in which large amounts of gas, including methane, are produced by an unusual fermentation in the rumen with consequent distension and sometimes death of the animal.

Effects of excess potassium

Effects of excess potassium result from its ready absorption by the plant and its mobility in the soil. As a result magnesium deficiency can be induced in crops and sometimes in stock. Clovers and potatoes are most seriously affected. Applications of excessive amounts of potassium in fertilizers are, therefore, not only uneconomical and wasteful but can have detrimental effects on the yield of crops.

Magnesium

Magnesium deficiency occurs mainly on soils such as

Podsols which have at one time been very acidic or on Calcareous soils in which there is a high ratio of calcium to magnesium.

Until the 1940s the soils of the British Isles were reasonably well supplied with available magnesium. Cropping demands were not high and farmyard manure rich in magnesium was widely used. Fertilizers were not 'purified' and the impure potassium minerals used in compound fertilizers contained appreciable amounts of magnesium. Grass/clover ley systems were widespread and the clover with its ability to extract more magnesium from soil than most species ensured a good turnover of available magnesium.

Since that time several changes in farming practice have greatly enhanced the risk of magnesium deficiency. An intensive liming campaign, in which mainly calciferous lime was used, started in 1938 and continued for more than 30 years. Magnesium has been virtually eliminated from fertilizers. Less farmyard manure is being used. Grass/clover leys have been largely replaced in many areas by pure grass swards. During the same period, crop yields have increased greatly and, with this, the demand for magnesium from the soil.

The combined result of all these factors has been a considerable increase in magnesium deficiency in soils, plants and animals.

Diagnosis of magnesium deficiency

Deficiency symptoms on plants are useful in diagnosis but are not specific in some crops. Expert opinion should be sought. In some cases, especially in cereal crops, symptoms are transient and disappear when the root system becomes established and explores greater volumes of soil.

In sharp contrast to potassium deficiency symptoms, the leaf margins of magnesium-deficient plants tend to stay green. Yellowing occurs in blotches between the major veins of broadleaved crops, the small veins being pervaded by yellow (Fig. 16.2). In cereals, symptoms take the form of roughly oval yellow patches 2–3 mm in length and crossing two to three veins. Brassica crops take on brilliant orange, yellow, red and purple colours but these are not specific to magnesium deficiency and can be caused by waterlogging, frost and other factors.

Soil analysis is not very helpful in predicting magnesium deficiency except in extreme cases. Plant analysis may be used to confirm plant symptoms. A further aid to diagnosis is the occurrence of hypomagnesaemia (low blood magnesium) in sheep and cattle, which can cause death through a tetanic condition. If this occurs where a high proportion of animal feed is from crops raised on the farm, it is a strong indication of a magnesium deficiency in the soil, which may produce herbage low enough in magnesium to affect the animal even though no plant symptoms are shown.

Prevention and correction

In acidic soils which need regular liming, the risk of magnesium deficiency can be reduced or prevented by the use of magnesian limestone instead of the more widely obtainable calciferous limes. The extra cost in areas remote from the sources is well worth while. It will usually be sufficient to use magnesian and 'ordinary' limestone alternately at similar rates of application. It is very important to understand that, although this will prevent magnesium deficiency in crops, it will not guarantee that stock will be free from hypomagnesaemia. Supplementary direct treatment of the animal may well be required.

Prevention of magnesium deficiency in soils which have a reserve of lime and do not require liming may be achieved by the use of neutral magnesium salts such as kieserite, $MgSO_4.H_2O$, applied in solid form, at rates of 300–600 kg/ha. Calcine magnesite, MgO, may also be used at rates of 150–300 kg/ha.

If magnesium deficiency symptoms occur on crops early in the season, the application of magnesian limestone at this stage would be completely ineffective because of its slow action. Epsom salts, $MgSO_47H_2O$, or the cheaper but less soluble kieserite can be applied as a spray on the leaves of the crop (4 kg/ha of Epsom salts in 200 litres of water *or* approximately half that amount of kieserite in the same amount of water). Alternatively, some 300–400 kg/ha of kieserite may be applied to the soil. There is no doubt,

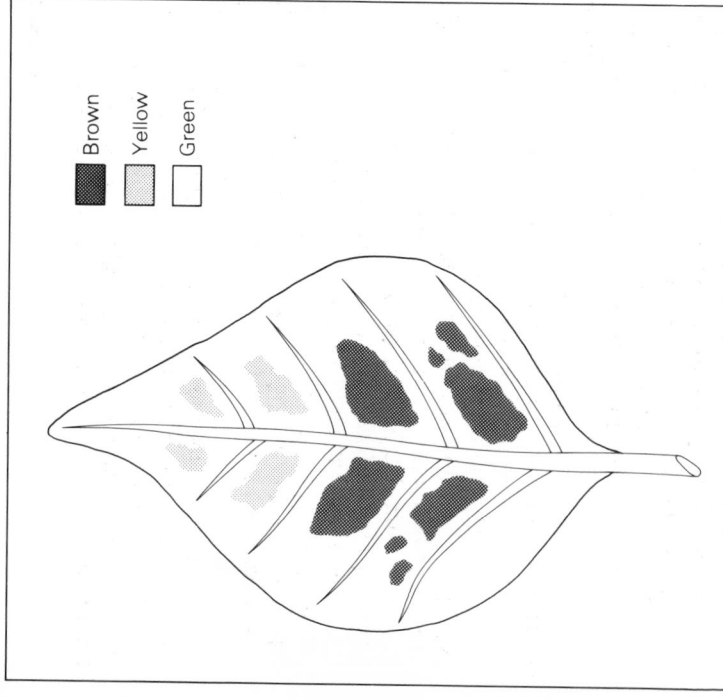

Figure 16.2 Symptoms of magnesium deficiency in potato

however, that the preventive measure of using magnesian limestone where practicable should ensure that no corrective measures would be needed.

Effects of excess magnesium

Adverse effects of excess magnesium are rare. They will never result from recommended applications of magnesian limestone or magnesium salts and are restricted in the British Isles to soils derived from ultra-basic rocks, rich in magnesium, in the Highlands of Scotland. These soils are infertile but the cause of this has not been firmly established.

Calcium

Calcium is an essential element but its direct functions in the plant take second place to the regulation of soil pH using limestones and chalk. Calcium deficiency can occur in some unlimed Peats and Podsols. Liming corrects this. Chapter 15 deals in detail with liming and the effects of excess calcium.

Sulphur

In the British Isles, sulphur deficiency was uncommon until recent years because of pollution caused by the burning of coal and other fossil fuels. To the lee of industrial urban areas, vast quantities of sulphur dioxide were released into the atmosphere, dissolved in rainwater and, thereby replenished soil sulphur.

During the first half of this century, much sulphur was also added to the soil in fertilizers. The most widely used nitrogen fertilizer at that time was ammonium sulphate, $(NH_4)_2SO_4$, which contains rather more sulphur than nitrogen. Superphosphate, then the most popular phosphate fertilizer, contains sulphur in the form of gypsum, $CaSO_4.2H_2O$. A standard application of compound fertilizer for potatoes, based on these substances could add to the soil fortuitously 100–130 kg of sulphur per hectare. Since 1950 ammonium sulphate and superphosphate have gradually been replaced in compound fertilizers by ammonium nitrate and ammonium phosphates, none of which contains sulphur.

The combined result of reduced pollution, changes in fertilizer composition and failure to recognize the role of sulphur as a major nutrient, is an inevitable and rapid decline in sulphur reserves. There are already signs of deficiencies in Eire and in parts of Scotland remote from industrial areas, particularly in high rainfall areas on light-textured soils, low in organic matter.

Diagnosis of sulphur deficiency

Crop symptoms are similar to those for nitrogen deficiency, pale green or yellow leaves and stunted plants. Soil analysis is at present of very limited value in predicting sulphur deficiency because of the uncertain contributions of mineralized organic sulphur to plant nutrition in a particular season. Analysis of plants at an early stage of growth can give some indication of the existence of deficiency conditions. Detailed analysis will reveal a reduction in total plant protein and also in the sulphur-containing amino-acids.

Prevention and correction

Now that the input of sulphur from pollution and compound fertilizers has been drastically reduced, virtually no measures are being taken to prevent sulphur deficiency.

The cheapest and most effective way to replace the sulphur which is being lost is by application of gypsum, $CaSO_4.2H_2O$, which can be obtained from natural sources in England. It is also a waste product in the manufacture of soluble phosphate fertilizers from rock phosphate. It is particularly unfortunate that this 'waste' gypsum is being dumped when 10 years from now it will be needed as a routine fertilizer in low-pollution areas.

The amounts of gypsum that would need to be applied

annually to the soil, simply to replace the sulphur taken off in crops would be some 80–160 kg/ha, or possibly even higher for brassica crops, which are rich in sulphur.

At present, negligible amounts of sulphur are applied. The main reason for this is, as with all developing deficiencies, a lack of appreciation of the scale and urgency of the problem. The onset of deficiency conditions will be gradual and the farmer cannot be blamed for not wishing to pay for and apply to the soil materials to which there will be no immediate response.

None the less widespread sulphur deficiency is inevitable within the next decade in the British Isles.

Effects of excess sulphur

The effects of excess sulphur usually occur, not within the soil, but as a result of sulphur dioxide descending on the leaves of plants in atmospheric pollution and scorching them.

Trace elements

Table 16.1 summarizes the rates of application of trace elements recommended for treating deficiencies, along with the most suitable chemicals and methods of application to

Table 16.1 Application of trace elements for the correction of deficiencies

Element	Chemical and rate of application	Method of application	Period of effectiveness	Precautions
Manganese	Manganese sulphate (MnSO$_4$.3H$_2$O) 4–6 kg/ha	Spray on to foliage immediately symptoms appear (or before if there is known to be a problem)	1 season	If applied in combination with herbicides, check compatibility
Boron	Borax (Na$_2$B$_4$O$_7$.10H$_2$O) 20–45 kg/ha	'Boronated' fertilizers applied for crops with high boron requirement	1 season	Avoid application to crops with low boron requirement, especially cereals
Copper	For soil application, copper sulphate (CuSO$_4$.5H$_2$O) 20–50 kg/ha	In solution to bare soil or in solid form to grassland	5–8 years	**Keep all stock off grassland until excess copper has been removed by rain**
	For foliar application to arable crops, copper oxychloride (CuOCl$_2$) 3–5 kg/ha	Spray on to foliage immediately symptoms appear	1 season	
Molybdenum	Sodium molybdate (39% Mo) or ammonium molybdate (54% Mo) 1.2–2.5 kg/ha	In solution to the seedbed	1 season	
Iron	Iron–EDTA chelates 0.5–1.0 kg iron per hectare	Spray on to foliage as soon symptoms appear	1 season	

Table 16.1 (contd.)

Element	Chemical and rate of application	Method of application	Period of effectiveness	Precautions
Zinc	Zinc sulphate ($ZnSO_4.H_2O$) 45–90 kg/ha	In solution to bare soil	3–5 years	
Cobalt	Cobalt sulphate ($CoSO_4.7H_2O$) 6 kg/ha	In solution to pasture	3–4 years	Consider cost/effectiveness compared with direct administration to stock

use. Reasons for the recommendations and precautions required are given in the sections on individual elements.

Manganese

Manganese deficiency is caused by the conversion of manganous compounds to the manganic form which is unavailable to plants. This is greatly encouraged by high pH conditions and good drainage. There are large quantities of manganese in many mineral soils and the deficiency is usually induced by liming. The exception to this is some deep Peat soils which contain very little total manganese.

Until the last decade it could be stated with confidence that the greatest risk of manganese deficiency occurred in sandy soils with high organic matter contents and pH values greater than 6.3, in dry seasons. Manganese deficiency was rare on heavy soils but in recent years there has been an increase of incidence, especially in barley crops, associated with poorly structured heavy soils. This is probably the result of the gradual ageing of oxidized manganese compounds following the great liming campaign of the 1940s and 1950s. If so, a progressive increase in manganese deficiency may be expected over the next 20 years.

Diagnosis of deficiency Manganese deficiency is most serious in cereal crops. It is worse in oats than in barley. Wheat is less affected. The yield of oats and barley crops can be greatly reduced unless the deficiency is corrected.

Other crops, including sugar beet and potatoes, are also affected.

Figure 16.3 illustrates manganese deficiency symptoms in sugar beet, oats and barley. Deficiency symptoms on the leaves are quite characteristic in some crops. In broadleaved species they take the form of an intensive mottling with very small yellow spots between the vascular network of veins while relatively small veins remain green. This is the main distinction between manganese and magnesium deficiency symptoms in which the yellowing pervades the smaller veins.

In barley, numerous dark brown or black lesions 1–2 mm long are found along the veins, giving a 'leopard spots' appearance. The symptoms in oats were among the earliest diagnosed trace element deficiencies. Compared with barley, the lesions are larger, 2–4 mm in length. They are greyish brown, giving rise to the name 'grey speck' and tend to be concentrated about one-third to half way up the leaf, spreading across the leaf and causing the collapse and folding down of the upper part of the leaf, which then dies. After the symptoms appear, cereal crops commonly take on a pale yellow, nitrogen-deficient appearance.

Symptoms of manganese deficiency are seldom found in seedlings or young plants because there is sufficient manganese in the parent seed to carry them through this stage. Even in severe cases, the first two leaves of cereal crops will not be affected, but the third one will.

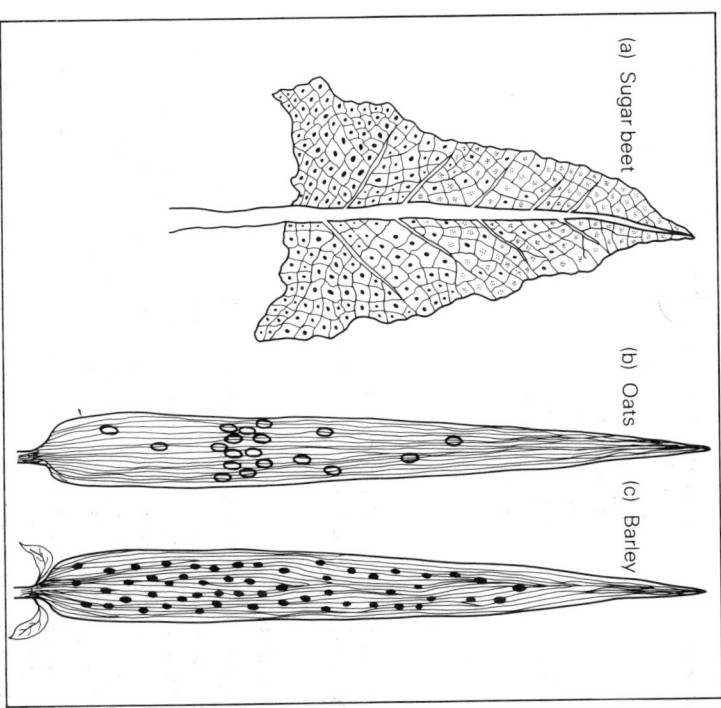

(a) Sugar beet   (b) Oats   (c) Barley

Figure 16.3 Symptoms of manganese deficiency in sugar beet, oats and barley

Also among the earliest trace element deficiencies to be identified in Britain was manganese deficiency in peas, 'marsh spot', so named because of its occurrence in the East Anglian fens. The symptoms cannot be seen until individual peas are removed from the pod, cut in half and found to have a dark spot 1–2 mm in diameter in the centre. The peas are inedible and, if used for seed, germinate very poorly.

Prevention and correction The only contructive preventive measure on acidic mineral soils is to maintain the pH below 6.5 by avoiding overliming. This is not always feasible if lime-loving crops like barley or sugar beet are to be grown and it is then necessary to take the risk of manganese deficiency and resort to corrective measures. There are no adequate preventive measures that can be used on Calcareous soils where the pH is permanently high or on deep peats in which the total manganese content is intrinsically low.

Crops showing manganese deficiency should never be treated with manganese salts applied to the soil. This is both costly and ineffective because the applied manganese is immediately subject to the same influences that have made the 'natural' manganese unavailable.

Instead, the crop should be sprayed with a solution of manganese sulphate, $MnSO_4.5H_2O$. Spraying is urgent if symptoms appear, despite the fact that rainfall sometimes brings about a recovery without the use of a spray. Responses to sprayed manganese sulphate can be quite spectacular. If the soil has a history of manganese deficiency preventive action can be taken by spraying as soon as there is sufficient foliage, irrespective of symptoms. The timing of manganese sprays for cereals coincides with that for herbicide treatments. If consideration is being given to a combined spray, the compatibility of the herbicide with manganese sulphate should be checked. Some herbicides form gelatinous precipitates if mixed with manganese sulphate and spray nozzles may be blocked.

There are some chelates, compounds of manganese with organic substances like ethylene diamine tetra-acetic acid, EDTA, which may be applied to the soil or used in sprays. They are expensive and are generally no more effective than manganese sulphate.

Effects of excess manganese Excess uptake of manganese by agricultural crops is usually associated with acidity in mineral soils (pH values less than 5.0). Plants grown on such soils can absorb toxic amounts of 500–1000 mg of manganese per kilogram of dry matter. Liming the soil to raise the pH into the 5.5–6.5 range will correct manganese toxicity. Overliming must be avoided. Symptoms of manganese toxicity are stunted plants, yellowing and death of leaf tips, and stubby brown roots.

Boron

Boron deficiency tends to occur on naturally acidic, low-humus soils and is made worse on such soils by liming, particularly overliming. It is prevalent in areas of low rainfall.

Diagnosis of boron deficiency Unlike other trace element deficiencies, boron deficiency seldom results in leaves showing symptoms. Diagnosis from the outward appearance of the plant is impossible in most crops because the symptoms are internal. Swedes grown for human consumption can reach the shops and be the subject of customer complaints before boron deficiency is identified. Root, brassica and legume crops are most susceptible. Symptoms are seldom found in cereals and grasses.

Figure 16.4 illustrates the symptoms in swedes, sugar beet and Brussels sprouts. The condition in swedes is known as 'heart rot' or 'raan' and in sugar beet as 'crown rot'. There is in each case a breakdown of cell structure and brown or black lesions develop. Secondary bacterial rots are common especially in swedes, whole fields of which can rot, creating an objectionable smell. Even mildly affected swede roots do not keep well and serious economic losses can occur. Boron-deficient sugar beet contains less sugar than normal plants.

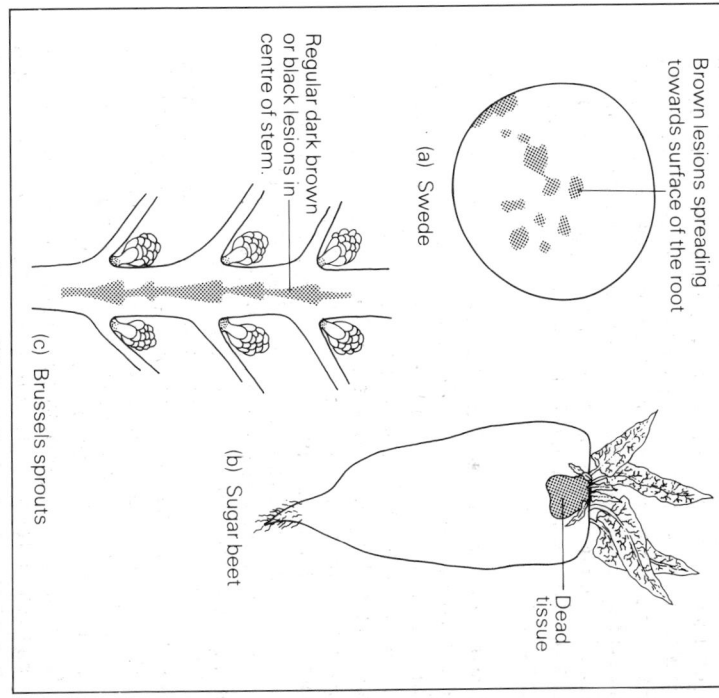

Brown lesions spreading towards surface of the root

(a) Swede

Regular dark brown or black lesions in centre of stem.

(c) Brussels sprouts

(b) Sugar beet

Dead tissue

Figure 16.4 Symptoms of boron deficiency in swedes, sugar beet and Brussels sprouts

Soil analysis gives a reasonably reliable guide to the risk of boron deficiency. Both soil and plant analysis can be used to confirm field diagnosis.

Prevention and correction   If soil analysis indicates low available boron, or if there is a history of deficiency symptoms in crops, susceptible crops should receive an application of borax, ($Na_2B_4O_7.10H_2O$) to the soil before sowing. It is critically important that applications should *not* be made for cereal crops. The margin between deficiency and toxicity of boron is very small and what would be a corrective treatment for swedes can be toxic to cereals. Borax cannot be applied as a spray because of its low solubility. The quantities required (Table 16.1) are so small that distribution in solid form, even if mixed with a diluent such as fine sand, is very difficult. However, fertilizer manufacturers can supply 'boronated' fertilizers in which borax is incorporated in the granules. Although costly, these fertilizers give excellent boron distribution.

Corrective measures, once symptoms have been seen in a crop, are not possible because the crop is usually maturing by that time. Some reduction in yield is inevitable. This is why preventive measures are essential.

Effects of excess boron   The margin between deficiency and excess is perhaps narrower for boron than for any other element. The position is made worse by the different requirements and tolerances of groups of crops. The most common cause of boron toxicity in agricultural crops is the accidental application of boronated fertilizer to non-tolerant crops, particularly cereals. This can happen if spare boronated fertilizers are stored over winter and 'used up' in the following season without an appreciation of the risks. The symptoms of boron toxicity in cereals are quite unique. In the seedling, no pigment production occurs and the leaves are white. The toxic effect may be lethal unless heavy rainfall dilutes the boron in the soil. The plant will produce new leaves, first with chlorophyll only near the tip and then completely green. Partial or total recovery will follow.

Copper

Copper deficiency can occur on a wide range of soils but is most likely to be found on Podsols, Peats and Calcareous soils. Long-term leaching of Podsols has resulted in copper deficiency and this is exacerbated by reclamation of these soils for arable agriculture because of the increased demand of subsequent crops for the element and because it becomes less available as a result of liming. Some Podsols contain so little *total* copper (less than 2 mg/kg) that they could not supply crops for more than a few years even if all the copper were in forms available to the plant. Shallow chalky soils (Rendzinas) are also subject to copper deficiency. In both types of soil problems are likely to arise in dry more so than in wet seasons. Copper deficiency was originally identified on newly reclaimed Peat soils and one of its names is 'reclamation disease'.

Light-textured soils of other types can develop copper deficiency as a result of overliming and drought.

Diagnosis of copper deficiency Cereal crops, except rye, are susceptible to copper deficiency and the symptoms are fairly characteristic. If the deficiency is severe the leaves die back from the tips and tend to spiral. Hence the name 'wither tip' is given to the disease. The leaves are very pale green and no grain is formed. The plant has a similar appearance to that resulting from wind-blast and the roots are so weak that the plants can be pulled out of the ground with ease. The symptoms which occur when deficiency is less severe are very frustrating to the farmer. The crop makes apparently normal vegetative growth and forms ears but no grain is produced. The term 'blind ear' is applied to this form of the disease. From a distance the crop appears normal but on harvesting can yield as little as 0.5 t/ha.

Symptoms on other crops are not so reliable. Peas fail to develop in the pod and shrivelling of leaf tips occurs in sugar beet and clover with some general yellowing. Defoliation and shoot die-back are found in top fruit crops.

A more complex diagnostic problem is found when animals suffer from copper deficiency. The disease, known as 'swayback', in ewes and young lambs causes paralysis and usually death of lambs. It is certainly associated with low copper soils but the availability of copper within the animal is also greatly influenced by its intake of sulphur and molybdenum. Both low blood copper and swayback can certainly occur when soil copper is adequate for crop growth.

Soil analysis is a useful guide to the adequacy of soil copper for crops and plant analysis can confirm deficiency.

Prevention and correction Prevention of copper deficiency in progressive agriculture is difficult because of the depressing effect of liming on copper availability, and also the increased requirement of crops.

Corrective treatments can be applied either to the soil or to the crop. For rapid results on the current crop a spray of copper oxychloride, $CuOCl_2$, should be made to the leaves. Copper sulphate, $CuSO_4.5H_2O$, is *not* recommended for this purpose because it can cause serious scorch on foliage.

Unfortunately, where crop symptoms occur, only partial recovery of the current crop can be expected. Long-term treatments are necessary. Copper sulphate is cheap and satisfactory for this purpose. To get even distribution it is preferable to spray the material in solution on to the bare soil, where possible. Copper sulphate solutions can, however, cause serious corrosion of metallic spray nozzles and other parts of the machinery and application is a job for the specialist. Alternatively, the material can be applied in solid form, but the rate of application is so low that there are distribution problems. It would be ideal to incorporate the required amounts of copper sulphate into granular compound fertilizers.

Applications to grassland should preferably be made as solid copper sulphate. If it is sprayed on the herbage, serious scorch can occur. Copper sulphate is very toxic to stock and, if an application has to be made to grassland, it is critically important to withdraw all stock from the area until there has been sufficient rain to remove excess copper from the herbage.

Effects of excess copper   There are no extensive areas of natural copper toxicity for crops in the British Isles, and the soil can absorb considerable quantities of copper without adverse effects on crops. There are areas where soil copper values are still high as a result of the regular use of Bordeaux mixture, a mixture of lime and copper sulphate, in the treatment of potato blight half a century ago, but crop yields on those areas seem normal.

It is sensible to avoid large applications of any trace element to soil and, therefore, sewage sludge known to be high in copper should be avoided or applied at very low rates. Slurry from pigs fed with copper-rich diets is also high in copper and should be applied in small quantities, avoiding grassland.

Molybdenum

The molybdenum content of British soils is generally low, apart from one or two small areas. The requirements of plants are also low. The main cause of molybdenum deficiency is soil acidity. The availability increases with pH so that deficiency is not common in well-limed agricultural soils.

Diagnosis of molybdenum deficiency   Symptoms are not easily recognized on most crop species. The older leaves become yellow prematurely and then become scorched but this is also symptomatic of other causes. However, on brassica crops, the symptoms are very striking. In cauliflower, for example, the leaves become very narrow, twisted and curled around an apparently normal midrib and the plant becomes blind, with no 'curd'.

Soil analysis can give some guidance on the risk of molybdenum deficiency and plant analysis can be used to confirm it. A secondary analysis can show a build-up of nitrate in leaf material because of the role of molybdenum in the reduction of nitrates in the plant.

Prevention and correction Maintenance of soil pH values in the range 6.0 to 6.5, by routine liming, is usually sufficient to prevent molybdenum deficiency. The only exception to this is in some Peats which have very low total molybdenum.

If deficiency does occur, applications of sodium or ammonium molybdate, should be made to the soil.

Effects of excess molybdenum Some poorly drained clay soils, with approximately neutral pH values and high total molybdenum content derived from black marine shale or other sources, produce pasture which is rich in molybdenum. The growth of the grass is satisfactory but the pasture is said to be 'teart' and animals feeding on it suffer from a disorder which responds to the feeding of copper in the diet. The reasons are not completely understood but the excess molybdenum seems to induce copper deficiency in the animal.

Iron

Iron is abundant in mineral soils. Whenever deficiency occurs it is induced by high pH, usually in Calcareous soils, and oxidizing conditions in the soil. Iron deficiency can occur in soils containing 2 per cent, or even 4 per cent of total iron. Conditions in which iron deficiency occurs are also conducive to manganese and magnesium deficiency and all may occur in the same plant.

Diagnosis of iron deficiency This deficiency is mainly a problem in top fruit orchards on chalk or limestone-derived soils. In horticulture it affects 'calcifuge' plants such as rhododendron. Plants suffering from iron deficiency do not form chlorophyll and the leaves are a pale 'sulphur' yellow with the small veins slightly greener. The condition is sometimes described as 'lime-induced chlorosis'.

Soil analysis and analysis of mature leaves can help in prediction and confirmation of field diagnosis.

Prevention and correction There is no way to prevent iron deficiency in Calcareous soils. The condition can be alleviated by the use of chelated iron compounds used in sprays.

Effects of excess iron Iron toxicity is rare in the normal pH range of agricultural soils, although it probably contributes, with manganese and aluminium, to the death of plants in acid soils.

Zinc

Although zinc deficiency is widespread in some parts of the

USA, no conclusive cases have been reported in British field crops.

Effects of excess zinc   Problems of excess of zinc in the British Isles occur in limited areas around old metal mining sites, where zinc toxicity has caused leaf yellowing and failure of cereal crops. Pollution with zinc from smelters, chemical plant, factories and sewage is the real hazard. Some sewage sludges contain vast amounts of zinc and should not be applied to the land.

Cobalt

Cobalt deficiency is associated with particular parent materials, predominantly Old Red Sandstone, Silurian and Ordovician shales, but has also been reported on other types of soil. Cobalt deficiency is, unfortunately, made worse by liming and drainage operations.

Diagnosis of cobalt deficiency   Although cobalt is essential for nitrogen fixation in leguminous crops, visual symptoms of deficiency in crops are not known in the British Isles. Lack of nodulation on the roots of subterranean clover and lucerne has been observed in cobalt-deficient areas of Australia and it may be that similar symptoms in wild white clover in some British hill pastures are associated with cobalt deficiency. Cobalt deficiency does not affect

non-leguminous crops.

Deficiency is best diagnosed by the occurrence of 'pine', a debilitating disease of sheep in which growth is retarded, appetite is lost and anaemia follows. It is caused by lack of vitamin B_{12} which contains cobalt in its molecular structure. Cattle may also be affected. Soil and plant analyses are useful for confirmation of deficiency.

Prevention and correction   Control of cobalt deficiency is best achieved by administering cobalt to the animal, either in a mineral mixture or as a 'bullet' of cobalt oxide, given orally, which releases cobalt slowly into the rumen.

Applications of cobalt sulphate, $CoSO_4.7H_2O$, can be made to pasture and will increase the cobalt content for some years. Amounts required are very small and must be applied either as a spray or incorporated in fertilizer. This has been done on a large scale in New Zealand for many years using 'cobaltized superphosphate'. At present the cost of cobalt is very high and applications to the hill and marginal pastures, where deficiency occurs, are prohibitively expensive.

Selenium

The causes of selenium deficiency are not yet well established, although it seems to be prevalent on strongly leached sandy soils. Light-textured soils that are deficient in copper and cobalt may well be deficient in selenium. The areas of selenium deficiency in the USA strongly suggest an association with Podsols and, in Australia, deficiency certainly occurs most in high rainfall areas.

Diagnosis of selenium deficiency There are no known symptoms in plants. Selenium deficiency in animals causes a form of muscular dystrophy, known as 'white muscle disease', mostly in grazing sheep and in calves. There is evidence that, even if this does not occur, growth and fertility of livestock can be adversely affected. The selenium requirement of stock is still not well established, but tentatively, values of less than 0.03 mg/kg of selenium in the diet dry matter is thought to indicate deficiency.

Prevention and correction In cases of suspected deficiency, selenium supplements should be fed to stock. The effects of applications of sodium selenite, Na_2SeO_3, to the soil are not well established.

Effects of excess selenium The marine black shales rich in molybdenum are also rich in selenium. Toxicity of selenium to stock can occur in areas where soils are derived from these rocks. Animals grazing herbage containing more than 5 mg/kg of selenium, in the dry matter, generally 'do badly' and can suffer from loss of hair, rough coats, and a sponginess and splaying of the hooves. This condition was known in central Eire long before its cause was established, and stock were shifted from selenium-rich pastures until their hooves had recovered to some extent. Imported diets, low in selenium, to supplement the intake from pasture, can be employed to prevent toxicity.

Other trace elements

There is some evidence that silicon, although not essential for most crops, plays some part in the nutrition of grasses and cereal crops, particularly rice. Deficiency of silicon is very unlikely in British soils.

There seem to be no immediate problems of deficiency of chlorine or chromium in plants.

Chlorine in the form of the chloride ion, Cl^-, is a major element in animal nutrition but the needs of animals must be met in mineral supplements and there is no point in adding more chloride to soils than that already added in fertilizers. The requirements of animals for fluorine, iodine and sodium must also be supplied, if necessary, in mineral supplements.

Comment

17

The reader will have found, in chapter after chapter, reference to the beneficial effects of organic matter in soil. The slow and insidious decline in the amount of humified organic matter in soils under arable agriculture is an important restricting factor to soil fertility, particularly in the drier parts of the British Isles. I am not, and never have been, a member of the 'Organic School' and I find many of their attacks on the use of fertilizers and other 'poisons' both misguided and detrimental to the production of sufficient food for the world population. None the less, I am in no doubt whatever that the maintenance or improvement of the humified organic matter content of our mineral soils is essential to long-term fertility. It will help greatly in the prevention of cultivation pans, surface compaction, frost heaving, clod formation, wind erosion, capping and puddling, drought, low-nutrient status and the excessive leaching of lime and fertilizers. It will also assist in the removal of excess water from the surface soil, the warming up of soils in the spring and will help to create a type of soil structure in which fine feeding roots can ramify in search of nutrients. This is a formidable list of benefits. The return of organic matter to the soil in crop residues and manures also replaces some of the nutrients removed by high-yield crops but there is an inevitable loss from the system in crops and animals sold off the farm. Therefore in the amount of organic matter that can be returned to the soil under any system of agriculture there are insufficient quantities of nutrients for total replacement. The deficiency must be made good by the application of fertilizers or by manures brought in from other sources.

The gradual nature of organic matter decline and failure to appreciate the need for conservation commonly leads to critically low levels of organic matter being reached before remedial action is taken. If these levels are reached, the way back is long and painful.

The key to organic matter conservation is in the attitude of the farmer and his advisers to crop residues, including cereal straw, which are too frequently regarded as a time-consuming nuisance and a difficult disposal problem. In fact, they are a valuable resource to be husbanded carefully. It has been amply demonstrated in Chapter 10 that the return of all possible crop residues to the soil, supplemented by processes such as green manuring that can be used without unnecessary delays to the production and handling of profitable crops, can maintain and increase soil organic matter within economic agriculture.

Another factor of the utmost importance is the difficulty of ensuring that the water supply to the plant is always sufficient but never excessive.

In the climate of the British Isles there are variations, locally, countrywide, from day to day and month to month. We have large tracts of land which are too wet and equally large areas which are too dry for maximum crop production. Less than one-third of our total land area has no serious problem one way or the other.

There are areas, not only those with excessive rainfall, where there are soils in which waterlogging restricts the yield of crops. There is a need to publicize and encourage the adoption of new techniques in the instalment and spacing of drains and the use of secondary treatments such as subsoiling, moling and the use of permeable fill. There is also a need for a greater appreciation of the damage that can be done to surface soil and drain lines by the use of very heavy machines to lay the drains.

Perhaps even more important, on a countrywide basis, is the need to provide, and where necessary subsidize, irrigation water and equipment, so that irrigation may be practised much more widely in those areas where drought (as estimated by current methods) occurs less than 5 years out of 10. The establishment of an adequate piped water supply would also remove the frustration of growers in areas of high drought risk in their quest for scarce water.

To do this effectively would involve major commitments on a national scale and there is little evidence that agricultural planners give priority to it.

There is no doubt that, in large areas of England and the dry east coast of Scotland, lack of water counteracts much careful husbandry and restricts the efficiency of fertilizer use more so than any other single factor and thereby restricts crop yields.

Soil acidity is not now a major limiting factor to crop production.

production. The benefits of liming are well understood but the present practices and methods of liming merit reconsideration.

There is certainly a good case for using magnesian limestone as a routine for alternate dressings. Even if a premium has to be paid for this material compared with calciferous limestones it is by far the cheapest way of adding magnesium, a major nutrient, to soils which are being depleted because of offtake in crops and lack of other magnesium supplementation in fertilizers.

Most of the fundamentally acid soils under agriculture have now been brought to a state in which only maintenance liming is required. We have become accustomed to and have accepted the system of liming at regular intervals of several years. This is convenient and is done mostly by contractors at very reasonable prices. There is, however, a strong case for reconsidering this method especially where intensive cropping is practised. Under many intensive cropping systems lime is exhausted rapidly and local surface acidity can occur especially with minimal cultivations. A workable alternative would be to apply lime on an annual basis in the same way as fertilizers and this might be done by using limestone ground more coarsely than at present. The production of computer models of the rate of lime loss in specific soils, under various systems of management, at different rates of nitrogen fertilizer application, would help greatly in assessing the amounts to be applied each year.

The trend over the last 20 years towards the use of heavier, more highly powered implements for cultivations, harvest and ancillary operations such as drainage seems to be continuing. There are undoubtedly advantages in time-saving but the use of these implements when soils are wetter than their plastic limits, and especially on heavy-textured soils, has led to a vicious circle of events. The heavy implements can cause compaction and pan formation. In order to correct this, still heavier implements are used for deep ploughing or subsoiling and further compaction can be caused, reducing the number of days when the soil is in a suitable state for cultivation and increasing the need for machinery which will do the job quicker. This equates with yet more highly powered machines with greater fuel requirements and greater potential for causing damage to the soil.

There has been some reaction against the use of heavy equipment and its costs by the use of minimum tillage or direct drilling and we are learning more every year about the applicability of these techniques to various types of soil. The methods have great attractions in terms of organic matter conservation and structure maintenance and will undoubtedly be more widely adopted within the next 10 years. There is a need, however, to define more closely the soil types and conditions for success and the economics of the systems.

Fertilizer technology is now very advanced and we have at our disposal a wide range of excellent compound NPK fertilizers as well as 'straight' fertilizers containing only one of these elements. The fertilizers available are now about as concentrated as they can be without resorting to the use of hazardous chemicals such as ammonia or elemental phosphorus and no further major changes in the composition of fertilizers can be expected.

The outstanding need in forecasting fertilizer requirements is to improve the rule-of-thumb methods at present used to decide upon rates of application of nitrogenous fertilizers. It is virtually certain that soil analysis will not help in this matter because of the unpredictable nature of the available nitrogen in the soil from day to day or week to week. The most promising approach would be a refinement of methods at present used to decide upon rates of application, using such factors as previous cropping, history of manuring and fertilizers, estimates of residues from leguminous crops, soil organic matter content, crop requirements and winter rainfall. The use of computer models would be helpful in such an exercise. The rewards of accurate forecasting of fertilizer nitrogen requirements of crops could be great in terms of increased crop production and avoidance of waste and pollution but the prediction will always be bedevilled by that great unpredictable, the British climate, and we shall never achieve accuracy.

It is tempting to speculate that some form of 'slow-release' or 'release-on-demand' nitrogen fertilizer might be developed but the present range of slow-release compounds are expensive and are dependent on the activities of soil micro-organisms for the speed at which their nitrogen becomes available.

The preoccupation with N, P and K in fertilizers, and the greater demands of crops for other essential elements resulting from steadily increasing yields have created stresses on the supplying power of soils for sulphur and the trace elements as well as the already mentioned need for magnesium. Within the next few years it will become necessary to apply sulphur compounds as part of the fertilizer routine. In some areas, unpolluted by sulphur from the atmosphere, it may already be rewarding to apply sulphur in the form of 'straight' calcium sulphate, gypsum, or to choose from the very restricted range of compound fertilizers containing potassium sulphate as their source of potassium. Such fertilizers have previously been used specially in the production of high quality potatoes and vegetable crops.

It is more difficult to envisage a way forward in meeting the demands of crops for trace elements. The use of 'shotgun' treatments, containing several trace elements, either in sprays or in fertilizers, is usually inefficient and can be hazardous because of the narrow margins between deficiency, sufficiency and toxicity for some elements. The only hope seems to be in the continued improvement of forecasting the risks of deficiencies of individual elements. Even if this is achieved the use of preventive rather than corrective measures will still be difficult because of the capacity of soils for immobilizing trace elements such as manganese and boron.

Finally, quite apart from these technical points, there are problems of communication between farmers, agricultural advisers and research and development workers in soil science, plant nutrition and animal nutrition.

These branches of science are not ends in themselves and if they appear difficult and esoteric to the layman this is a major fault on the part of the scientist.

Although I hope that this book will be read by scientists it has been written with the minimum of jargon to stimulate the interest and to meet the needs of farmers, agricultural students, advisers and anyone else concerned with the prosperity of the agricultural industry. If it helps them to understand and care for the soil and to communicate their problems to the scientist in language that all of them can understand it will have served its purpose.

Further reading

General

E. W. Russell (1980) *Soil Conditions and Plant Growth* (12th edn). Longman
N. C. Brady (1974) *The Nature and Properties of Soils* (8th edn). Collier Macmillan (New York); Macmillan (London)

Physical problems

Soil physical conditions and crop production (1975) Ministry of Agriculture, Fisheries and Food, Technical Bulletin 29. HMSO

Information on local soils

Memoirs and other publications of the Soil Survey of England and Wales; *Memoirs* of the Soil Survey of Scotland. HMSO
(These are available for various parts of UK but not for the whole area.)

Soil formation, classification and colour photographs of soils

E. A. Fitzpatrick (1980) *Soils, their Formation, Classification and Distribution.* Longman

Index of definitions and descriptions

General index

plastic pipe drains, 146, 152
plasticity
  clay, 72
  effect of excess water, 158
  organic matter, 67, 169–70,
    172
  plastic limit, 130, 169–70, 172
poaching, 99, 125, 150, 173
Podsols, 31–3
  fertility, 180–2
  gleyed, 34
  reclamation, 196–7
pore space, 88, 98–9, 168
potassium, 118–20, 206–8
  availability, 118–20
  deficiency, 206–8
  excess and magnesium
    deficiency, 208
  fixation by clays, 118
  mineral reserves, 118, 119
  pathways in soils and plants,
    118–19
profile, soil, 23–7, 32–7, 124–9
puddling, 86, 165
puffy seed beds, 164–5

raised beach soils, 19
reclamation of acid soils, 196–7

saltation, 175
sampling
  plant, 183
  soil, 183–5

sand
  fine, importance of, 73
sedimentary soils, 20–3
sedimentary rocks, 22–3
selenium, 223
silt
  accumulation in drains, 147, 149,
    150
slurry, 137
soil analysis, 185–6, 203, 205, 208,
    209, 212, 218
  rapid testing, 186
soil conditioners, 142
soil association, 46–7
soil formation, 6–28
  processes, 8–20, 23–7
  relation to soil fertility, 27–8,
    180–2
soil groups
  agricultural potential, 45
  Brown Earths, 30–1, 44, 45
  Brown soils, 44, 45
  Calcareous soils, 38–9
  Chernozems, 44, 45
  Chestnut soils, 44, 45
  Gleys, 33–7
  Great, 42, 44–5
  in the British Isles, 29–41
  Intergrades, 39–41
  Peats, 37–9
  Podsols, 31–3, 44, 45
  Prairie soils, 44, 45
  Red Desert soils, 44, 45

soil profile, 25–6
  choice of site, 124–5
  colour, 32, 34–7, 67, 126–8
  development, 23–7
  field examination, 126–9
  horizons, 24, 26, 32–3, 34–6,
    127–8
  mature, 24
  pans, 126, 129
  pits, 126–9
  root patterns, 127–8
soil sampling, 183–5
soil series purity, 50
Soil Survey, 45–56
  derived maps, 50–6
  maps, 45–50
  memoirs, 47
soil temperature
  effect of soil colour, 67
  effect of water content, 158
soil variation, 39–41, 183–4
soil water deficit, 159
stones
  effects on soil properties,
    76–7
  shape and size, 13
structure, 28, 70, 78–87, 134–42
  classification, 83, 84–5
  development, 78–82
  effect on availability of nutrients,
    109–11
  effect on availability of water, 92,
    93–5

effect on irrigation needs, 163
factors affecting, 82–3, 127–8,
  140–2
  good, 83
  of calcareous clays, 171–2
  of horizons, 79, 127–8
  poor, 83
  stability, 83–7, 130–1
  types, 78–86, 93–5, 128, 169–70
structure maintenance, 64, 134–42
  direct drilling, 141
  drainage, 141
  gypsum, 142
  lime, 142
  minimum disturbance, 140–1
  organic matter, 134
  soil conditioners, 142
  subsoiling, 141
  timing of cultivations, 141
structure type
  angular blocky, 80, 84, 94, 128
  crumb, 81, 82, 84, 128
  granular, 82, 84, 128
  massive, 85, 128
  platy, 83, 85, 86, 94, 128, 169–70
  prismatic, 78, 79, 84–5, 93–5, 128
  single grain, 78, 84
  sub-angular blocky, 80, 81, 84,
    128
subsoiler, 148, 156
subsoiling
  conditions for, 156–7
  effect on structure, 141